# THE
# HISTORICITY
## *of* NATURE

WOLFHART PANNENBERG

# THE
# HISTORICITY
# *of* NATURE

*Essays on Science and Theology*

EDITED BY
NIELS HENRIK GREGERSEN

TEMPLETON FOUNDATION PRESS
*West Conshohocken, Pennsylvania*

Templeton Foundation Press
300 Conshohocken State Road, Suite 670
West Conshohocken, PA 19428
www.templetonpress.org

Designed and typeset by Kachergis Book Design

Templeton Foundation Press helps intellectual leaders and others learn about
science research on aspects of realities, invisible and intangible. Spiritual realities include unlimited love, accelerating creativity, worship, and the benefits of
purpose in persons and in the cosmos.

LIBRARY OF CONGRESS CATALOGING-IN-PUBLICATION DATA
Pannenberg, Wolfhart, 1928–
The historicity of nature : essays on science and theology /
Wolfhart Pannenberg ; edited by Niels Henrik Gregersen.
p. cm.
Includes bibliographical references and index.
ISBN-13: 978-1-59947-125-9 (pbk. : alk. paper)
ISBN-10: 1-59947-125-6 (pbk. : alk. paper)   1. Religion and science.   I. Title.
BL240.3.P36 2007
261.5´5—dc22
2007013251

Printed in the United States of America
08 09 10 11 12 13   10 9 8 7 6 5 4 3 2 1

# Contents

# CONTENTS

# Introduction

## Wolfhart Pannenberg's Contributions
## to Theology and Science

### BY NIELS HENRIK GREGERSEN

The title of this collection of essays, *The Historicity of Nature*, encapsulates Wolfhart Pannenberg's fundamental thesis of the historical character of nature. Not only is human existence shaped by historical decisions and cultural turning points, but nature also has an irreversible history in which virtually every event is unique and new complex structures are continually built up.

## NATURE'S HISTORICITY AND
## THE LAWS OF NATURE

Many people will be prepared to accept the thesis of nature's historicity—at least, concerted developments in twentieth-century science point in this direction. At the macroscopic level, the big bang theory gives evidence that our cosmos has had a unique history since the outburst of energy in the early beginnings of our universe. At the microscopic level, quantum mechanics suggests that nature is like a decision-making machine that randomly produces new quantum events; each singular event is contingent in its emergence, even though collectives of quantum events statistically fall within the scope of the Schrödinger equation. Finally, according to our best knowledge, historicity also characterizes the medium-sized world in which we have our living. The second law of thermodynamics, formulated in 1865, stated that the same energy can't be used twice to perform the same work. Statistically, differences in energy density and temperature will even out over time, so that entropy overall increases. In local settings, however, living systems are still able to use the remaining energy differences from the influx of energy and build up their own structures. With this background, the German physicist and philosopher Carl Friedrich von Weizsäcker argued in an influential book, *Die Geschichte der Natur* (1948), that the second law entails the very principle of the historicity of nature: "No event can be exactly repeated. Nature is an irreversible process."[1] Von Weizsächer's book strongly influenced Pannenberg's early work in science and religion. Later, the importance of the second law for nature's historicity was confirmed by the work of the Belgian chemist Ilya Prigogine and colleagues, who pointed out that the second law also facilitates bifurcation processes in chemical and biological systems far from thermodynamic equilibrium, some with long-ranging evolutionary effects. Finally, current evolutionary theory shows the extent to which chance or contingency prompt evolution to develop in diverse directions, reducible neither to physical laws of nature nor to putative "designs."

Pannenberg's thesis of nature's historicity, however, is more radical than so far described. According to Pannenberg, not only the phenomena of nature are historical in nature but also the very laws of nature. Alternatively, we could imagine that all concrete phenomena were in flux, whereas the laws of nature remain stable and are able to determine the future course of natural events. In this scenario, the surprising novelty of natural phenomena is a mere appearance. Admittedly, since epistemic unpredictability does not itself warrant an ontological indeterminism, such deterministic view of nature remains a residual theoretical possibility. However, the argument for such a view is based on ignorance (what we do not know) rather than on scientific knowledge (what we actually do know).

Pannenberg offers a principled argument against the idea of laws being fixed and all-determining. The application of any law of nature always requires a certain state of affairs in order for the law to cover the phenomena. Mendel's law of inheritance, for example, requires the emergence of living beings with sexual reproduction. Like species, laws of nature also emerge in history and may go extinct. Even very fundamental laws of physics (such as gravity or electromagnetism) require the existence of material stuff differentiated in space, a situation that may not have been the case in the first moments of our known universe and may not be the case forever.

Thus, the radical nature of Pannenberg's thesis of the historicity of our universe comes to the fore in his view on laws of nature. During the 1960s, he developed a specific version of what is usually termed the *regularity view* of laws of nature: laws of nature are not like Platonic entities that pre-exist the material world and prescribe in detail what happens next; rather, they describe in a post-hoc manner the regularities observed. Based on this knowledge, laws of nature hypothesize what happens next, assuming that all other things are equal (*ceteris paribus*). Given that things are not always equal, no law, and no ensemble of laws, can describe reality exhaustively. In Pannenberg's view, the individual occurrences, in all their endless varieties, are prior to the causal connections (subsequently formulated in laws of nature) that arise out of the interplay between individuals.

Pannenberg also offers a corresponding theological interpretation of the regularity view. According to the biblical view, creation is not a completed affair but something ongoing. God is seen as constantly at work as Creator. But variation, not repetition, is the characteristic of creativity. As such, God is both the creative source of individuality and the faithful provider of the regularities that result in the universe at large. For Pannenberg, our understanding of the universe as a unified whole should embrace, rather than delete, the sense of individuality.

Readers well versed in the current debate between Darwinian thinking and the Intelligent Design movement may find the following statement surprising: "[I]n a theological interpretation of nature, the element of chance or contingency is even more important than design, because contingency and the emergence of novelty correspond to the biblical view of God's continuously creative activity in the course of history and in the world of nature" (this volume, 46). In fact, Pannenberg questions the overuse of the metaphors of purpose and design in theology because these metaphors easily slip into anthropomorphism. God is imagined like a human calculator, who first conceives of a plan and then executes this plan in strategic acts, where something is treated as means to ends. As eternal, however, God must be equally close to all creatures and to any moment in time: past, present, or future.

In this context, Pannenberg proposes to speak of God as acting in the world as the "power of the future." A God who has already pre-conceived a plan in the past and impresses it on the present can only be seen as a determining force, while the God who is coming to the pres-ent moment out of what is not yet realized facilitates and calls forth the freedom of the creatures. Therefore, a central point of Pannenberg's theology of evolution is to speak of God as acting in the world of cre-ation in such manner that "spontaneous self-organization," the princi-ple of the superabundance of natural processes and of human freedom, can take place. However, the God of futurity must be equally close to all creatures. As the eternal future, God is, has been, and will be the facilitating source of the life of past, present, and future generations.

In this manner, Pannenberg views the world of creation as histori-cal from beginning to end and across its dimensions. There is no abso-lute divide between nature and culture, human freedom and the laws of nature, as was the argument in the Enlightenment tradition following upon René Descartes (1596–1650) and Immanuel Kant (1724–1804). A separation cannot be maintained between a tidily ordered nature forever condemned to self-repetition and a human culture marked by freedom and flux. As we are now going to see, Pannenberg early on aimed to overcome this nature–culture divide.

## FROM THEOLOGY OF HISTORY
## TO THEOLOGY OF NATURE

In the 1950s, Wolfhart Pannenberg (born 1928) was already working on "a theology of universal history," which he later expanded into his theology of nature. After having completed two highly specialized dis-sertations as a young man, in 1953 and 1955, Pannenberg was gener-ally recognized as a rising star. Yet he was also viewed with some suspi-cion by fellow theologians. Looking back to the atmosphere of German Protestant theology in the 1950s and 1960s, the young Pannenberg (himself a Lutheran) was swimming against several currents. At a time when liberal Protestants argued that the age of metaphysics and dogma was gone, Pannenberg studied medieval thinkers (the topic of his dissertations). At a time when existentialism reigned, he was a theologian with comprehensive intellectual ambitions that could best be described as metaphysical. When neoorthodox theologians argued that the message of the Bible and the Church was different from all other religions and worldviews, Pannenberg resisted the isolation of the Christian tradition from other traditions. At a time when many said that Christianity at its core is not really a religion but a secularizing force, Pannenberg saw Christianity as one religion among others. In a climate when a high degree of minimalist cleanness was observed in theology (backed up by the then-existing hegemony of Protestantism in Northern Europe and North America), Pannenberg interpreted Christianity as the most syncretistic religion that had ever seen the day's light. In an era, in which many theologians depicted Judaism and other religions as "religions of the law," Pannenberg claimed that Christianity remains highly dependent upon its Jewish resources. Indeed, according to Pannenberg, the scope of Christianity can be evaluated only in the light of a comprehensive theology of all world religions.

As early as 1961, Pannenberg and a circle of former Heidelberg students issued a book with the programmatic title *Revelation as History*. According to Pannenberg's brief explanation, the group "had not intended to cause a theological revolution, but only wanted to provide a more solid biblical foundation for a key concept of theology, the concept of revelation."[2] The tone of the book was indeed calm and objective, but it was a manifesto that challenged basic assumptions of both existentialism (Rudolf Bultmann) and neoorthodoxy (Karl Barth). The English Macmillan edition aptly displayed in large type on the front cover: "A proposal for a more open, less authoritarian view of an important theological concept."

Bultmann (1884–1976) and Barth (1886–1968)—the two giants of their generation of German theologians—shared the view that the proper object of theology is "the Word of God." In their view, divine self-revelation takes place in the biblical or preached Word, not in the public realms of history or nature. The historical figure of Jesus is irrelevant, as are empirical features of the external world; what matters is the preached Word that elicits faith in its hearers: "the Christ-event."

Uneasy about such abstract theological argument, Pannenberg

argued that, in the biblical tradition, it was not words of God that were considered to reveal God but the actions of God in the midst of ordinary history. Revelation is "no secret or mysterious happening," nor is it anything ostentatiously divine, like a theophany. Rather, the revelation of God takes place *indirectly* in historical accidents: "Historical revelation is open to anyone who has eyes to see. It has a universal character."[3] Yet, as events among other events, acts of revelation need interpretative eyes in order to discern the concurrent patterns of divine agency. Just like nature and culture should not be contrasted, so should objectivity and subjectivity not be pitted against one another in the concept of revelation. What happens in history has meaning, and the task of theological interpretation is exactly to explicate the meaning of happenings and their constellations.

Soon, Pannenberg pursued this program within a comprehensive theological framework. These endeavours led at first forward to his Christology, *Jesus—God and Man*, published in 1964. Later, it was carried forward in many publications, finally to be completed and synthesized in his three-volume *opus magnum*, *Systematic Theology*, published 1988–1993.

This introduction is not the place to discuss Pannenberg's theological project in its entirety. But one might wonder why Pannenberg, in the midst of his intense theological work with classical, medieval, and modern resources, began to take serious interest in the sciences. Pannenberg's own answer is clear and concise. Typically for his mindset, he does not refer to stories about commitment and personal preference but to the inner logic of faith: *if* Christians (along with Jews and Muslims) confess God as the Creator of all that exists, *then* there *cannot* be two separate domains—one explained by the sciences and another addressed by theology and the human sciences.

To begin with, Pannenberg devoted himself primarily to the understanding of human nature in discussion with proponents of philosophical anthropology and disciplines such as biology and ethology. This led to the publication in 1962 of a short and concise book, *What is Man?*,[4] later in 1983 followed up by the major work *Anthropology in Theological Perspective*. Pannenberg's method in this work is to accept secular descriptions of humanity as basic but also as provisional descriptions of the reality of human beings.[5] Descriptions of humanity within philosophy and the sciences may not be complete, especially not with respect to its religious dimensions. Theologically important dimensions of humanity are easily overlooked in disciplines such as sociology, psychology, or cultural anthropology, sometimes for solid methodological reasons, sometimes as a result of scholarly bigotry. Hence, there is both room for and a need for specific theological contributions to the understanding of humanity. The reader of *Anthropology in Theological Perspective* will, therefore, discover that Pannenberg's interaction with scientists often involves dispute as well as consonance. The theological task of the dialogue is for Pannenberg not to reach an external, apologetic accommodation; theologians should also be involved in developing fresh interpretations out of theology's own resources. The theologian has the task both to listen and speak, both to take and to give. This dialectical approach is operative also in the essays of this volume.

As early as 1962, however, Pannenberg engaged with a small group of scientists and theologians in Karlsruhe, who deliberately addressed the possibilities of a theology of nature (at that time considered a dubious project). He later continued these discussions on the interface between physics, fundamental ontology, and theology in Heidelberg and in Munich. Finally, in 1970, he copublished with the physicist A. M. Klaus Müller his first publication on theology and the natural sciences. Here Pannenberg presented his view of the historicity of the laws of nature: laws are grounded in contingent events, rather than contingent events being exceptions from laws.[6] The essays grouped in Part Two of this volume primarily address the relations between physics and theology.

Alongside the new engagement with fundamental theories of physics, Pannenberg now began to work intensely with issues concerning the philosophy of science. His major book *Theology and the Philosophy of Science*, published in 1973,[7] shows that Pannenberg found it pertinent to develop a methodological self-critique within theology as well as in the human sciences in general. The first two essays of this volume present Pannenberg's most basic convictions on these difficult questions. While reading these chapters, the American reader should be alert to the fact that the German term *Wissenschaft* is used more generously in German than in English. The concept of *Wissenschaft* covers the human and social sciences (*Geisteswissenschaften*) as well as the natural sciences (*Naturwissenschaften*). Also the term for *religious studies* is in German *Religionswissenschaft*, that is, "science of the religions."

Bearing these particularities in mind, it is characteristic that Pannenberg not only challenges the Kantian divide between humanity and nature but also the principled contrast between the natural and the human sciences. What in Descartes was an ontological contrast between the realm of the body and the realm of the soul was in the nineteenth century transposed into a methodological divide between the physical and the moral sciences (J. S. Mill). In Germany, Wilhelm Dilthey (1833–1911) further developed the distinction between *Naturwissenschaft* and *Geisteswissenschaft* into a theory of essentially different methodological tasks of *Erklären* (scientific explanation) and *Verstehen* (hermeneutical understanding). Quite a few scholars up to this day, in particular within theology and the human sciences, still hope to uphold this dichotomy. But it was a central concern of Pannenberg to overcome this split by demanding that theological interpretations also be analytically intelligible in terms of content and testable in terms of truth-claims. Theology should no longer be able to evade rational control and shelter itself against scientific critique.

It should be noted, however, that Pannenberg also argues the other way around: the natural sciences should no longer be able to claim that they represent an objective knowledge of pure facts. All scientific theories are "underdetermined by data," that is, they say more about reality than can be empirically tested. In addition, no scientific theory can express itself without being accompanied by philosophical interpretations. Just as there exists no pure theology, there exist no pure sciences.

## ON THE AIMS AND ESSAYS OF THIS VOLUME

The present collection of essays has three interrelated aims. The first is to make available for English readers Pannenberg's distinctive essays on the interface between Christian theology and the natural sciences.

Some of these articles already exist in various English-language jour-
nals and volumes; others are translated for the first time into English
or have not been published before. In this respect, the present volume
supplements an earlier collection from 1993 of Pannenberg's essays,
*Toward a Theology of Nature. Essays on Science and Faith*, edited by
Ted Peters. A second goal for this publication is to show how Pannenberg has been in constant dialogue not only with physics and biology
but also with the sciences of cultural anthropology, sociology, and psychology. These disciplines are, of course, quintessential when it comes
to understanding the characteristics of human nature, regarding both
its positive and destructive elements. A third aim is to document Pannenberg's intellectual discussions with some of strongest proposals of
American philosophy and theology, in particular, the heritage of Paul
Tillich and Langdon Gilkey and of Alfred North Whitehead and process theologian John B. Cobb Jr.

Part One entails two essays on the methodology of theology that
summarize Pannenberg's views on methodological status of theology.
These chapters show how strongly Pannenberg advocates the classic
view that nothing less than God can be the ultimate object of theology. Since most religions claim to reflect experiences of the Divine, religions are not taken seriously unless theology combines a proximate
focus on the religious experiences of humankind with a philosophical
investigation of their ultimate truth-claims. What in "lived" religion is
a life-experiment, from day to day tested by first-hand experience, is,
at theological level, treated as a thought-experiment to be tested by a
hypothetical reasoning in face of scientific knowledge and philosophical reflection. "A religious proclamation of a particular God can be
acceptable only if it plausibly explains how the world as we know it,
including human society, emerged from that god" (this volume, 6), says
Pannenberg. Pannenberg, thus, distances himself from self-immunizing
ways of interpreting theology. However, Pannenberg is also a theologian who does not excuse himself for being a theologian. The task of a
Christian theology is no less than to rethink the Christian tradition in
its universal scope. Exactly because God, according to the biblical traditions, has revealed himself in the public realms of history and nature,
theology cannot insulate itself behind holy texts and church doctrines,

nor hide itself behind havens of piety. The Bible, the Church, and the Christian faith do not only exist *in* the world; they are also speaking *about* the same world as the sciences do. In this sense, a Christian theology must be able to explain itself in the context of scientific thinking as well as in the context of a theology of world religions.

What, then, about the differences between studying science and practicing religion? One seminal difference is that, whereas the sciences aim to explain empirical phenomena by theoretical unification, religions are concerned with the ultimate horizons of reality, including life-orienting questions about how to attune oneself to overarching patterns of meaning. This transempirical orientation of religion comes to the fore in the idea of God. In the monotheistic traditions, God is not thought to be an empirical object, as if God "existed" as one item among others (within the world or beyond the world). Rather, God is assumed to be real and effective by being the creative source, which informs, pervades, and surrounds everything that exists. "No one has ever seen God," as the Christian tradition soberly acknowledges (1 John 4:12). The reason is not that God is an absent reality but that God is the encompassing reality, the one "in whom we live and move and have our being" (Acts 17:28).

According to Pannenberg, theology in the proper sense should, therefore, not be seen as a purely empirical research program that solely aims to explain the data of religious humankind. Theology (so I would paraphrase Pannenberg's method) begins in a hermeneutical inquiry into the universes of meaning of religious traditions (experiences, texts, behaviors, etc.). Only when the intentions of religious texts and life forms are interpreted appropriately and the religious candidates for truth analyzed correspondingly can ideas of God be tested. There exist no direct God experiments, though. Nonetheless, the hypothesis of the pervasive reality of God can and should be *indirectly* tested, that is, via the *implications* of religious truth-claims for worldly realities.

The essays in Part Two deal primarily with the relationship between modern physics and problems of philosophical theology, in particular the God–world relation. Pannenberg observes that disciplines of science and theology only seldom meet directly because the languages are too different. However, they can meet in the *mediation of philosophical reflection* on issues of worldview and religious concerns. When scientists go public and inform the public about their methods and findings, they do not simply speak the language of mathematics; rather, they engage in a wider philosophical reflection about the meaning and implications of their scientific results for matters of worldview. Similarly, theologians do not just quote biblical texts or bits of liturgy. Rather, they present concerns of religious ideas and discuss their implications for understanding reality as a whole, which again involves philosophical self-reflection. In addition, the *history of the sciences* offers a platform for interaction between scientists and theologians. Pannenberg reminds us that even core scientific ideas such as "atoms," "energy," "fields" and "laws of nature" have religious or philosophical origins that continue to guide the self-interpretation of the scientists, wittingly or unwittingly.[8] Intellectual history may also identify scientific challenges to theology that are otherwise forgotten. In medieval theology, it was assumed that bodily movements are always moved by something. But after the principle of inertia was introduced by Descartes (even as nature's first law), continuous movement is seen as intrinsic to material bodies, without need of an external mover. As Pannenberg observes, "the Christian doctrine of providence never recovered from this blow, and very rarely has the issue been faced" (this volume, 59).

The discussions on physics and theology in Part Two gravitate around three themes on which Pannenberg has done innovative work. One topic is the relation between historical contingencies and laws of nature, a topic already touched upon. A second recurring theme is the reinterpretation of God's eternity and omnipresence in relation to relativistic time-space; a third is the concept of field as a foundational concept in physics as well as in theology.

The very notion of God as spirit (John 4:24) seems to suggest a fundamental opposition between spirit and matter. However, Pannenberg here points to the fact that, in the Hebrew Bible and early Christianity, the idea of spirit (*pneuma*, in Greek) was not originally conceived in a Platonic vein, as thought or mind (*nus*, in Greek). Both in Genesis (1:2) and in the Gospel of John (3:8), the divine Spirit is conceived of as a creative field operative over and between creatures. This view is closer to the notion of *tónos*, or creative tension, in Stoic monistic philosophy than to the dualistic orientation of Platonic thinking. According to the historian of science Max Jammer, the Stoic notion of a pneumatic field was a forerunner of the modern scientific uses of field concepts. It was Michael Faraday (1791–1867), in his reflections on electromagnetic phenomena, who reintroduced the concept of a "field of force" as a foundational concept in physics. The most basic physical reality is no longer the existence of separate bodies in inertial motion but the field which only occasionally manifests itself in discrete bodies. This view led to the later work of Albert Einstein's theory of relativity, according to which mass is a property of energy; most of energy, however, is not manifest as material particles but appears field-like. In *The Evolution of Physics*, Einstein and Leopold Infeld stated their position as follows: "Matter is where the concentration of energy is great, field where the concentration of energy is small."[9]

Responding to his critics, Pannenberg admits that his use of the concept of field to designate the operative presence of the Spirit of God in the world of matter indeed is metaphorical, but it is a metaphor that has gained a conceptual status in theology as well as in physics. Pannenberg does not argue that divine presence in creation in any way competes with the field-entities of physics, be it classical, relativistic, or quantum. But on theological premises, God must be at work "in and beyond the forces of nature without being exhaustively expressed by them" (this volume, 37). The theological concept of divine Spirit is both connected to and distinct from the concepts of field forces in physics.

A similar mode of argument can be found in Pannenberg's proposal to rethink the divine attributes of omnipresence and eternity. Pannenberg's philosophical theology is based on the notion of infinity: God is not an entity existing alongside other entities, as if God begins on the other side of created reality. God is rather the ultimate reality in and beyond all created reality. As such, God is the all-encompassing field who gives to the creatures a relative freedom. Hence, the omnipresent God must be thought to be constitutive of space, without God being identified with space (neither with the classic Euclidean space, nor with the geometry of relativistic time-space). With Isaac Newton (1642–1727), Pannenberg speaks of the immensity of God as being prior to the measurable space. Similarly, eternity must embrace time, without deleting the uniqueness of the individual moments. In line with a long tradition, Pannenberg understands eternity as copresent with all times: past, present, and future. But this concept of eternity is again reconceived in terms of the futurity: "Through the future, eternity enters into time" (this volume, 35).

Part Three addresses themes of anthropology and religion and brings together essays less known by English readers. A short essay (chapter 8) introduces Pannenberg's view on the discussion between the idea of creation and Darwinian evolution. The full importance of Darwinian theory for Pannenberg's thinking, however, can only be ascertained by studying his *Systematic Theology*, volume 2.[10] Here can be found extensive and detailed discussions on evolution, in which Pannenberg uses resources of evolutionary biology to emphasize interplay between irreducible novelty and the overall drive toward higher-order complexity in the history of creation. Also here, Pannenberg wants to balance the sense of the individuality with the universal features of reality.

Pannenberg's engagement with cultural anthropology and psychology is exemplified in this volume by the translation of central articles that address human religion, human spirit, and the human propensity toward destructive and self-destructive behavior. Pannenberg observes that several strategies are used to exclude the theme of human religion from treatment in disciplines such as cultural anthropology and sociology. Some simply tone down the centrality of religion to human beings; some declare religion to be illusory; others treat religion as an intermediary phase of human development with some secular utilities later to be taken over by secular functions. Pannenberg argues that this supersessionist approach to religion is unfounded. Human nature is always carried out between a self-centeredness based on needs and a self-transcending openness toward ultimate horizons of reality, including suprahuman agencies.

History testifies to that. Archaeologists use the findings of careful burials of bodies, with heads placed towards the east, as evidence of the humanity of extinct lines of hominids. Moreover, in critical discussion with Jean Piaget and Ernst Cassirer, Pannenberg advances the hypothesis that the cult may have been the origin of the acquisition and development of language. Ritual play, the expression of sounds, the coordination of movements, and the attunement to symbolic patterns of meaning have facilitated the transition from the language of imperatives (prescribing what participants should do) to the language of indicatives (describing what is), which includes references to symbolic realities in abstract terms. Language arises out of the religious emotions in such ritual plays. Accordingly, human history does not start with stable selves. An "I" presupposes a "we," and the sense of symbiotic union with reality (originally the mother or child carer) continues to inform the basic trust, without which adults do not dare to open up themselves toward ever-new groups and ever-new environments. Such limitless trust can find its anchorage only in the encompassing reality of a loving God.

Hereby, Pannenberg does not mean to present an argument for the reality of God. He wants to argue, however, that humanity cannot be understood when cut off from its religious roots. Both the stabilization of selfhood and the legitimacy of social forms require a religious dimension that transcends the needs and drives of individual egos. This also applies to the notion of human dignity. In cases of infancy, serious handicaps, or dementia, the infinite value of human persons can be maintained only if the human being is seen as sharing in the sanctity of God. The biblical notion of humanity as created in the "image and likeness of God" (Gen 1:26–27) exactly means that the dignity of the human person is derived from his or her continuing relation to God, rather than to intrinsic qualities.

Hence, the human spirit should also be seen as something more than a consciousness or self-awareness. The concept of "field" also has relevance for the understanding of the human being as a spiritual being, rather than as a lonesome soul. Just as the origins of language emerge within the spiritual field of ritual coordination, so does the field of human spirit ground the emergence of human subjectivity and self-awareness. The Kantian point of departure in a self-constituting transcendental ego can't be maintained philosophically, nor can it stand the test of science. We don't start out as stable selves; selves are developing over time. Personhood may, in this context, be seen as the provisional realization in the present moment of the genuine self that lies before the individual in his or her future destiny.

Chapter 11 offers a highly original analysis of the roots of human self-destruction. This essay discusses the theological notion of human sin in relation to phenomena such as drives and frustrations, fear and anxiety, depression and aggression. I commend this essay in particular because it offers a highly original—and as far as I'm aware—unparalleled analysis of the human urge toward destruction and self-destruction. The overall assumption is that human emotional life is more deep-seated and versatile than suggested by a purely mechanical model of instinctual drives. Neither can the urge to destroy be derived from externally imposed frustrations of society. Augustine (353–430) was correct in stating that there is here something fathomless and vulnerable in the human constitution of selfhood. The precarious and practically impossible balance between our constitutional self-centeredness and our constitutional openness drives human beings into aggression toward others or into the self-aggression of depression. These questions are worked out in detailed discussions with Augustine, Kierkegaard, Nietzsche, Freud, and with current scientific theories of aggression. The article also includes the recommendation that theologians revise tendencies within the Christian tradition (especially in Augustine's theory of original sin) that favor either self-aggression or aggression toward others.

Part Four brings essays that address Pannenberg's theory of meaning and his critical dialogue with the metaphysical approaches of Paul Tillich and Alfred North Whitehead and their contemporary interpreters.[11] Just as scientists aim to discover reality rather than construct reality, so Pannenberg argues that we do not only construct meaning but also discover a meaning that is already given prior to our attention. Certainly, our perception of the semantic meaning in texts or occurrences involves our full attention, but the process of finding meaning cannot appropriately be called solely our activity. In attention, receptivity precedes activity. Moreover, in our emotional life, we do not only experience isolated data; a wider whole of reality is present to us, partly in memory, partly in anticipation. Religion means that we experience particular events as carved out, as it were, from an infinite whole that we cannot encompass but only intuit in broad shapes, based on the fragments of our experiences. From this background,

Pannenberg argues that the meaning of our lives always lies in front of us, in our futurity.

These questions reappear in more technical form in Pannenberg's detailed critique of Whitehead's version of process philosophy. Pannenberg brings forward at least four serious criticisms of Whitehead's philosophy, the first three philosophical in nature, the last theological in nature. First, Whitehead's atomistic philosophy is one-sided by not giving equal status to the individual events and the broader forms of reality, such as fields in physics, organisms in biology, and the perception of meaning in human experience; these phenomena cannot be derived from the point-like events of Whitehead's "actual occasions." Second, Whitehead has in an unduly manner generalized William James' theory of momentary experience by presupposing that subjectivity exists at all levels of reality, from God down to atomic events. Third, even though Whitehead occasionally speaks about anticipatory feelings in actual occasions, he seriously constrains the idea of anticipation, since all events constitute themselves in the present moment, so that anticipation mainly refers to the objective influence of present decisions on the future situation ("objective immortality," as Whitehead called it). Fourth, Pannenberg argues that Whitehead's metaphysics needs far-reaching revisions in order to serve as a matrix for a Christian theology, revisions more radical than suggested by Gilkey or admitted by Cobb. Whitehead situates the source of creativity in matter rather than in God; hereby God is confined to a secondary and accompanying role in evolution by being the one hosting the ideal forms and proposing ways to proceed for self-determining creatures. But because God is not the one who creates out of nothing in the beginning, God cannot create new life at the end of creation. Whitehead, therefore, has no room for an eschatological completion of creation. Therefore, process theology either ends in a tragic view of life or in cherishing utopian hopes of an idealistic type, as it was the case in John B. Cobb's early theology.

Utopianism—either in Cobb's humanistic version (chapter 15) or in Frank Tipler's hopes for a digitalized future (chapter 16)—shows the extent to which anticipations of the future are an inescapable part of human imagination. The present volume, therefore, ends in questions of an eschatological nature, even though elaborate answers will not be provided here. Interested readers are advised to look into Pannenberg's *Systematic Theology*, vol. 3.[12]

## ACKNOWLEDGMENTS

I am grateful to Professor Wolfhart Pannenberg for trusting me to edit this volume and for agreeing to the proposed selection and ordering of articles. My thanks also extend to Linda Maloney for her fine translations of the German essays. Last but not least, I thank Templeton Foundation Press for supporting the publication of these essays and to Natalie Silver for her careful editorial work. I am in special debt to its Associate Publisher Laura Barrett for her constant commitment and pace.

## NOTES

1. C. F. von Weizsäcker, *Die Geschichte der Natur. Zwölf Vorlesungen* (Göttingen: Vandenhoeck & Ruprecht, 1979), 37.

2. Wolfhart Pannenberg, "An Intellectual Pilgrimage," *Dialog: A Journal of Theology* 45, no. 2 (2006): 188.

3. Wolfhart Pannenberg ed., *Revelation as History*, trans. David Granskou (London: Macmillan, 1968), 135. German original: *Offenbarung als Geschichte* (Göttingen: Vandenhoeck & Ruprecht, 1961).

4. Wolfhart Pannenberg, *What Is Man?*, trans. Duane Priebe (Philadelphia: Fortress Press, 1972). German original: *Was ist der Mensch? Die Anthropologie der Gegenwart im Lichte der Theologie* (Göttingen: Vandenhoeck & Ruprecht, 1962).

5. Wolfhart Pannenberg, *Anthropology in Theological Perspective*, trans. Matthew J. O'Connell (Philadelphia: Westminster Press, 1985), 59. German original: *Die Anthropologie in theologischer Perspektive* (Göttingen: Vandenhoeck & Ruprecht, 1983).

6. A.M. Klaus Müller and Wolfhart Pannenberg, *Erwägungen zu einer Theologie der Natur* (Gütersloh: Gütersloher Verlagshaus, 1970), 6. Pannenberg's article in this volume, "Kontingenz und Naturgesetz" has been translated into English in *Toward a Theology of Nature. Essays on Science and Faith*, ed. Ted Peters (Louisville: Westminster/John Knox Press, 1993), 72–122.

7. Wolfhart Pannenberg, *Theology and the Philosophy of Science*, trans. F. McDonagh (Philadelphia: Westminster Press, 1976); German original: *Wissenschaftstheorie und Theologie* (Frankfurt am Main: Suhrkamp Verlag, 1973).

8. See Wolfhart Pannenberg, "Contributions from Systematic Theology,"

in *The Oxford Handbook of Religion and Science*, eds. Philip Clayton and Zachary Simpson (Oxford: Oxford University Press, 2006), 359–71.

9. Albert Einstein and Leopold Infeld, *The Evolution of Physics* (New York: Simon and Shuster, 1938), 242.

10. Wolfhart Pannenberg, *Systematic Theology*, vol. 2, trans. Geoffrey W. Bromiley (Grand Rapids, MI: William B. Eerdmans, 1994), 115–35. German original: *Systematische Theologie Band II* (Göttingen: Vandenhoeck & Ruprecht, 1991).

11. See also chapter 5, which discusses Tillich and Whitehead in relation to Langdon Gilkey's work.

12. Wolfhart Pannenberg, *Systematic Theology*, vol. 3, trans. Geoffrey W. Bromiley (Grand Rapids, MI: William B. Eerdmans, 1998), chap. 15. German original: *Systematische Theologie Band II* (Göttingen: Vandenhoeck & Ruprecht, 1993).

# METHODOLOGY

## ✦ I ✦

# Theology Examines Its Status
# and Methodology

IN THE EARLY HISTORY of European universities, Christian theology was conceived in terms of a science—not only as divine wisdom (*sapientia*), but also as a science. Beginning in the thirteenth century, theology became a discipline of methodical argument, comparable
to other sciences in the line of the then-recently discovered Aristotelian
concept of science, with the difference that theology's principles are
not known to the human intellect by nature, but only through revelation. A science that argued strictly from its principles, however, would
proceed by deduction. But few theologians proposed to do so. Thomas
Aquinas, for example, in his *Summa Theologica,* did not argue directly
from principles of revelation (the articles of faith) but constructed a
systematic argument best described by his own term as an argument
by *congruence*: Such an argument does not proceed directly by drawing conclusions from principles but inversely from data shown to be
consonant with and leading to the principles. In this way, Aquinas first
argued for the existence of God from the effects of divine activity in
the world, and from this point of departure, he reconstructed the affirmations of the Christian doctrine step by step. Early Protestant dogmatics followed a similar procedure, though in the place of principles,
they had the biblical scriptures.

By the early eighteenth century, natural theology as a starting point was replaced by the concept of religion as the source of common knowledge about God, although this knowledge still had to be specified by revealed religion. The main representative of this viewpoint was Friedrich Schleiermacher. But to take religion as starting point for theology meant to make anthropology the basis of theology because religion is a human experience and activity. In this case, would one ever arrive at a concept of God as proper subject of theology, who could claim priority over everything else? In reaction to this argument, in the early twentieth century, a theocentric turn occurred, espoused first by Erich Schaeder and then by Karl Barth. In substance, this argument is quite persuasive. If one takes the task of theology seriously, it must be theocentric, not anthropocentric. But then the question is whether one can get to the theocentric position simply by way of a personal decision, as Barth originally proposed. In that case theology is still anthropocentric because it is based on a subjective decision. The problem is not substantially eased, when the starting point is taken not with the individual decision of the theologian, but with the church's proclamation of the word of God, as in Barth's mature dogmatics. The authority of the church, based on the scripture, once more takes the place of a presupposition of theological argument. The acceptance of that presupposition, however, remains a matter of individual decision. Therefore, it is preferable to reach the theocentric position by argument. In this sense, it seems inescapable to pick up the thread of the argument where modernity has left it, with religion as a human phenomenon. This position should also underlie the approach of philosophical theology in the modern situation, which corresponds to the fact that in antiquity religion existed before philosophy and was presupposed in the origin of the philosophers' critical quest for the true nature of divine reality. Starting the argument, however, with religion as a human phenomenon must not mean that religion is *only* a human phenomenon. Such a belief would forgo the intention and truth-claim of religion itself. Religions usually assign the origin of their teachings to God or to the gods. Taking religions seriously, then, not only requires a phenomenology of religion on the basis of human experience and institutions but an investigation of their truth-claims in talking about the divine reality. In this sense, a critical theology of religion is called for.

But how is this possible? It seems necessary to reconstruct rationally the truth-claims of a particular religion so that an educated judgment can be formed on the question: To what extent does the God of a particular religion illuminate to the satisfaction of its adherents, at least, the meaning structure of human experience with regard to the world of nature as well as to history and human society in such a way as to make plausible the claim that this God is indeed not only the creative origin of the world and of human life but also the redeemer from evil that human beings experience? To answer this question, Christian theology has to examine the truth-claim of the Christian proclamation of the trinitarian God as the creator and redeemer of the world, of both the world of nature and of human history.

This task first requires a hermeneutical investigation into the biblical roots of the Christian message: Is the trinitarian God identical with the creator God of the Hebrew Bible, the God of the covenant with Israel, whose coming kingdom Jesus proclaimed? This is the more traditional aspect of the theological task. An affirmative answer to the question largely depends on the exploration of Jesus' relationship to the God he proclaimed as his father and the father of all creatures, a relationship meant to include all human beings in their relationship to God the father. The method here is that of historical and hermeneutical investigation focused on the concept of God, not on whatever significance biblical texts may have in abstraction from their relationship to the God of Israel and of the Christian faith. Furthermore, implications in the teaching and behavior of Jesus and in Israel's witness to its God must be made explicit, since there is no overtly trinitarian teaching in the Bible. The implications of Jesus' proclamation about his own person and the significance of his cross and resurrection concerning his relationship to the God of Israel are of crucial importance in this argument.

A second requirement of the theological task is to clarify the concept of truth. Some would think that the task of theology is already accomplished when the trinitarian doctrine of the church, affirming the preexistence of Jesus and his divine Sonship, is shown to be in accordance with the scriptures. Such a view presupposes the divine authority of the scriptures, and the truth of Christian doctrine depends on being in agreement with the witness of the Bible. This is indeed an important aspect of Christianity's truth-claim, but more is required: The divine authority of the scriptures has to be argued for, not merely presupposed. This can be done only if the God to whom scripture bears witness is truly God, the Creator of heaven and earth and of every human being, not only in the perspective of a particular faith but with plausibility for all, at least in principle, in the sense that such plausibility may be claimed and argued for with good reasons. Therefore, Christian theologians have been concerned from early times that their teaching about God accord with philosophical monotheism, although this policy does not preclude critical argument against some philosophies. Agreement with philosophical monotheism means that the God of the Christians is identical with the one God to whom all human beings are related, and thus the biblical scriptures also may be accepted as issuing from that one God.

The philosophical teaching regarding the one God emerged in Greece in the sixth century BCE from a critical examination of the mythical tradition; this examination was guided by the question of how the divine origin of the cosmos must be conceived so as to become understandable as origin of the cosmos as in fact the cosmos presents itself to human experience. A religious proclamation of a particular god can be acceptable only if it plausibly explains how the world as we know it, including human society, emerged from that god. The nature of God must be conceived in such a way as to make it thinkable that the cosmos, as we know it, originated from this source.

One of the most important results of this critical examination of the mythical religious tradition was that the divine origin of the cosmos must be one because only on this condition is or can the unity of the cosmos be explained. In making the experience of the cosmos a criterion for the true conception of God, the Greek philosophers did not set up a criterion that was foreign to the mythical tradition because that tradition itself claimed that its gods were the origin of everything or at least of particular realms of reality. The philosophers raised the question whether it is indeed the case that such a god could be understood to be the origin of the world as we know it, and the result was that the mythical conceptions of the gods had to be radically revised. Thus, we see that the truth-claims of any religious tradition are judged not by some alien standard but by the implications of its own teachings.

When the Christian theologians incorporated philosophical monotheism in their argument for the universal truth of the biblical teaching about God and about the Christian trinitarian doctrine, they accepted that these beliefs must be consonant with our knowledge of the world in all its aspects, so that the claim that this God is the creator of the cosmos becomes plausible. This stance implied a coherence theory of truth: Christian teaching must be coherent with all aspects of the reality of the world and of human life: Only under that condition can the Christian claim that the biblical God is the creator of this world and of all humanity be plausible.

The truth-claim of the Christian teaching about the one true God, the creator of the universe, then, implies the acceptance of coherence as criterion of truth. But how can theology hope to satisfy this criterion? How can the claim to coherence with all sorts of human experience of the world ever be tested or confirmed? My answer is that this can be done in the form of a systematic presentation of the Christian doctrine of God, of creation, and of human history in terms of a history of salvation. Such a systematic presentation must be not only consistent within itself and consonant with the biblical witness but also coherent with regard to all matters that have to be taken into account in such a presentation. Each aspect must be reinterpreted when considered in its relation to God because the secular treatment of nature at large and of human nature abstracts from the relationship to God the creator.

This need for reinterpretation applies to data from human sciences and from history as well as to data from the natural sciences. Such reinterpretation must not do violence to the data, of course, but must be appropriate. The theological reinterpretation must be argued reasonably and plausibly and provide a systematic reinterpretation of all that is claimed to be God's creation. The systematic form of presentation corresponds to the requirement of coherence as a criterion of truth, standing as a testimony to truth insofar as it explores and presents the coherence of all issues concerned with affirming the trinitarian God as creator of the universe. In attempting such a systematic presentation Christian theology offers its examination of the truth-claim involved in the affirmation, that the trinitarian God is indeed the creator and redeemer of the world and of all humankind.

Of course, such an examination and presentation could never amount to a definitive demonstration of the truth of the Christian faith in the trinitarian God. The reinterpretation of the data from biblical exegesis, from human sciences, and from the natural sciences can be questioned, and the appropriateness of such a reinterpretation may be denied. Therefore, each attempt at the systematic presentation of the Christian doctrine remains a *hypothetical* reconstruction of the universal coherence of the phenomena of God's creation with the theological assumptions concerning the Creator. The hypothetical form of theological affirmations does not impair, however, their truth-claim. It belongs to the logic of affirmations in general, that upon reflection they present themselves as hypothetical: With each affirmation the question arises whether it affirms what is actually the case. Without this possibility of doubt, of questioning, we cannot have an affirmation or assertion. Therefore, the assertive intention of statements and their hypothetical status on a level of reflection or critical reaction are not contradictory. It is a condition of taking an affirmation seriously that it is possible to raise the question of whether the claim of the affirmation accords with reality. In all the sciences, humility is required: Our affirmations are not by themselves infallible but may be questioned. Theology is no exception to this rule. Nor does this fact impair the certitude of faith, a certitude that reaches beyond the hypothetical status of our affirmations, but may itself be tested. In my opinion, the admission that religious and theological announcements are hypothetical in their logical form, though strongly assertive in their pronouncement, is a condition of being taken seriously with their truth-claims.

What has been said so far on theological method refers specifically to systematic or dogmatic theology. Among the other theological disciplines, biblical exegesis and the study of church history apply the generally accepted rules of historical investigation and of hermeneutics. Ideally, these disciplines should also critically reflect upon and examine the secular formulation and use of those rules. They too need to reinterpret the relationship of history to God. Usually this task is left to the systematic theologian. In the case of pastoral theology, again, a critical reinterpretation of the data from relevant anthropological disciplines is necessary and may be done in cooperation with the systematic treatment of anthropology in the context of dogmatics. The focus of reflection on theological method, therefore, has to be on systematic theology. That discipline is not only concerned with the doctrine of God in the narrow sense—with the trinity, christology, and the doctrine of Creation—but also considers human sinfulness, redemption, the foundation and reality of the church and of the sacraments, and the life of the Christian believer. Finally, it investigates the basis and content of Christian eschatological expectation. All these matters, however, are related to the central doctrines concerning God and his revelation in Jesus Christ; thus, in some way, they belong to their explication, and the appropriate procedure is similar: Exploration of the biblical basis and of the general plausibility of the affirmations of Christian teaching as these questions have been treated in the history of Christian thought and as they should be restated at present.

In concluding these remarks, I may add some observations on how theological statements compare to those of some of the other sciences. One of the differences from the natural sciences is that theological statements cannot be tested by experiment because they do not affirm rules concerning recurring and repeatable sequences of events. Only hypotheses of law can be tested in such a way. The test of theological affirmations is rather in their hermeneutic correctness and systematic presentation. Such a presentation shows whether or not a particular affirmation is consonant with all pertinent data that have to be taken into consideration. In this regard, theological statements and their critical examination resemble the procedures in other hermeneutical disciplines like the study of history, where a course of events is reconstructed with due attention to all the available evidence. Sometimes the hermeneutical disciplines or human sciences have been opposed in principle to the natural sciences and to their procedures. I am not persuaded that this position is correct because I think that the natural sciences, in designing their theories, also aim at a form of systematic description and interpretation. They merely use other means to achieve that purpose because they are concerned for the repeatable patterns or laws that govern recurrent and repeatable processes, while the historical disciplines are concerned for unique and unrepeatable processes and phenomena, though historical processes may also comprise repeatable patterns. In the case of theology, the history of salvation—from the first act of creation to the final consummation of the universe—is just such a unique and unrepeatable process, though it certainly comprises the emergence of regularities, some of which may be described in terms of natural laws.

# Is There Any Truth in God-Talk?

*The Problem of Theological Statements from the
Perspective of Philosophy of Science*

IN THE *Theologische Aufsätze* published in honor of Karl Barth's fif-
tieth birthday in 1936, Heinrich Scholz addressed the question
"What is a theological statement?" His reflections on that question
referred to a discussion between Barth and him a few years earlier on
the possibility of Protestant theology as a science. In a lecture given
in 1931 in Bonn where Barth was then teaching,[1] Scholz had formulated, in a manner that is still instructive today, the minimal conditions any science or scholarly discipline must fulfill. He distinguished
between contested and uncontested minimum demands that must be
posed to any science. Barth responded in the first volume of his *Kirch-
liche Dogmatik* (1932), and his response amounted to an utter rejection.
Scholz's first demand, postulating the noncontradictoriness of every scientific statement, could not claim unreserved validity in theology. Barth
stamped the further requirements of unity of the object field—verifiability of statements, agreement with other sciences, and independence of
prior judgments—as "unacceptable": "Not an iota can be yielded here
without betraying theology, for any concession here involves surren-
dering the theme of theology."[2] In his essay on the concept of a theological statement, Scholz resumed the conversation with Barth, though

without any express reference to the previous discussion. In doing this, he attempted to show that, on the contrary, theology could not escape the demands of logic as simply as Barth had contended. In this article, Scholz understands a statement as a proposition that can be either false or true. The specifically theological character of a statement is said to depend on establishing or agreeing upon what is to be considered specifically theological. Scholz agrees with Barth's opinion that a theological statement is a statement about God—an opinion that surely coincides with the dominant view throughout the history of Christian theology. But the problem with this assertion is that not every statement about God is a theological statement in Karl Barth's sense. According to Scholz, Barth regards statements about God as theological only if they are not part of a rational theology. But the sense of this delineation cannot be clearly determined unless one is willing to define rational theology as the theology of the natural human being and a natural human being, in turn, as one "who is not enlightened by a theological statement in Karl Barth's sense." Thereby Scholz shows the absurdity of the idea that a theological statement is a nonrational statement about God. A theological statement, like any other, cannot evade its logical implications: "no one can sensibly forbid even a theological statement to have some kind of logical consequences." Hence, Scholz wished to "be allowed to support the proposition that the demonstrability of a statement as such may not be introduced to characterize the statement as not theological."

Karl Barth did not address Heinrich Scholz's arguments again. For Barth, after all, the irrational engagement of faith was the "risk" of what he called "altogether unsecured obedience" to the Word of God, which is only discernible as the Word of God in this very obedience.[3] According to Barth, all theological argumentation presupposes this act. And that was not only Barth's way of thinking, but Rudolf Bultmann's as well and likewise that of most of the prominent theologians of their time. So William Bartley could speak of a "retreat to commitment" (the English title of his book that appeared in German in 1962 as *Flucht ins Engagement*) as the characteristic of the whole of current Protestant theology: By appealing to the commitment of faith, theological thinking ultimately escaped any rational critique. This position was justified by the assertion that all thought rests ultimately on unprovable assumptions. Bartley shows that this assertion was indeed true of many forms of rational thought insofar as they attempt to ground themselves in certain ultimate certainties. Only a hypothetical and conjectural thinking, without any appeal to supposed ultimate certitudes, can be maintained against such objections. But precisely such thinking in hypotheses and conjectures is what characterizes modern science and its specific rationality.

The logical structure of every proposition implies a hypothetical character. To the extent that a proposition can be true or false and it is not yet determined which is the case, every proposition has the character of a hypothesis about reality. But the hypothetical character of propositions provides for the possibility of asking whether they are true or false, and thus of testing them. A proposition that, in principle, could not be tested to see whether it is true or false would not be a proposition; it would make no sense as a proposition. In this case, logical positivism is correct. But it has inadmissibly restricted the testing to which propositions may be subjected to a particular kind of test, namely, sense observation. To this it must be said that there can be other kinds of tests. Nevertheless, it remains correct that a proposition that, in principle, cannot be tested cannot be a proposition. The meaning of such a statement must, then, be something entirely different.

Hence it is not surprising that, in view of theological rejection of the susceptibility of theological propositions and their truth-claims to rational testing, one might come to the conclusion that theological statements are not propositions at all, not cognitive statements, but—as has been said—performative utterances: statements that do not assert anything about a matter that exists independently of them but rather express what is happening in the utterance of the statement itself. When, in response to an offer, I say "I accept it," by that very utterance, I accept the offer. When the pastor says, "I baptize you ...," through that very statement (and the action that accompanies it), he or she baptizes. People have attempted to understand statements like "I believe in God the Father almighty, the creator" similarly. In that case, this statement would contain no proposition about the existence of God or God's attributes but only the expression of a commitment, that of faith. But if we look closer, it is evident that the meaning of such a statement does contain a propositional element to which the commitment is linked. If there is no God, the statement "I believe in God the Father almighty, the creator" makes no sense. No commitment, no matter how emphatic, can counteract that fact. Therefore, every faith statement contains a core that has the character not merely of a performative utterance but also of a cognitive statement: the nature, that is, of a proposition. But as proposition, theological talk about God must be accessible to rational examination, for that belongs to the structure of propositions as such. And, as Scholz demonstrated, one ought not to avoid the logical consequences of what one says.

Thus, the postulates from philosophy of science that Scholz had formulated in his earlier essay as minimum demands to be addressed to every science are grounded in the logical structure of propositions: The requirement of *noncontradiction* arises from the fact that every proposition asserts something as true and thus excludes falsehood. The requirement of an *openness to being tested* is based on the truth that every proposition represents a hypothesis about an object, and that hypothesis may correspond or not correspond to the object—it may be true or false. These two demands are connected to the third, that science must address a *unified object* field, which is in itself distinguishable from statements about it. This distinction of the object from what is said about it arises from the hypothetical character of propositions: One of the most important services of language is, in one and the same act, to distinguish an object that can only be grasped in language from the form of its linguistic presentation. Therefore, a number of propositions can be understood to apply to the same object. The unity of the object or object field of a science thus constitutes the objective correlate and condition for the demand for noncontradiction among statements about this object. Therefore, the three minimum theoretical scientific requirements identified by Scholz as undisputed only make explicit what is already always contained in the logical structure of propositions. For this reason, Scholz was able to concentrate his argument with Barth on the question of the logical structure of a theological statement in their second confrontation about the possibility of a scientific theology.

The implications, coming from philosophy of science, of the logical form of a knowledge statement or proposition are uncomfortable for theology. This is evident from Barth's reaction, his rejection of Scholz's postulates as "unacceptable." But one may refuse to acknowledge the implications of one's own speech only at the price of refusing to know what one is saying. If theology is not to refuse the self-criticism that is associated with self-knowledge, it must subject itself to Scholz's challenge because these demands make clear something that has already been tacitly acknowledged by the fact that theologians do present propositions. The demand of noncontradiction can most easily be met. This demand does not mean that divine reality itself can be presented in theology as noncontradictory but only that the presentation itself must be subject to the discipline of noncontradiction. It cannot be excluded from the outset that unresolved difficulties in the subject can be described as such without involving the description itself in contradictions.

The second demand—the unity of an object field, which can be distinguished from what is said about it—presents significantly greater difficulties for theology. If, for example, a common self-definition of theology takes the Word of God as its proper object, it appears that the object cannot be clearly distinguished from statements about it: Even if theology presupposes that the Word of God is enacted in preaching or through the authority of the church, still the theologian, as a believer, must always decide that what is found in the preaching or the church's teaching or the Bible is the Word of God and not only human words. Even if one regards not the Word of God but God's very self as the object of theology, there is the difficulty of how God is accessible as a reality distinguishable from the propositions of theologians. That the reality of God does not appear to be distinguishable from the propositions of believers and theologians but instead appears in the form of such propositions and that, therefore, such propositions are not to be regarded seriously as propositions but appear to be fictions composed by the faithful and the theologians—that is the crisis of the God question in our times.

At this point, the question of theology's object, or rather whether theology has an object at all, shifts immediately to the question of the possibility of testing theological propositions, particularly their verifiability or susceptibility to being checked as regards their truth-claim.

This is the most difficult challenge to theology but one that it cannot avoid by any sort of protestations about its superiority or the unprovable character of divine truth. Such protestations may immunize theological discourse against critique but, at the same time, render it absurd because propositions that are, in principle, closed to any kind of critical questioning of their truth-claims are not propositions at all and hence can no longer be seriously regarded as propositions.

But how can it be at all possible to test theological statements? Propositions about God or about God's action or words or revelation—such propositions obviously cannot be tested against their immediate object. First of all, the very reality of God is disputed, and second, it would be contradictory to his divinity, as the all-determining reality, if he were to make himself available to human beings as a finite reality, reproducible at any time and at will, so that human propositions could be measured against him. God's reality—whatever else it might be—is not accessible in this way. Propositions about God, his actions and revelations, can, therefore, not be tested directly against their object. But that does not say that they cannot be tested at all. One can, after all, test propositions in terms of their implications. Propositions about divine reality or divine action can be tested according to their implications for our understanding of finite reality, insofar, that is, as God is asserted to be *the all-determining reality*.

Certainly, the notion of an all-determining reality is far from giving an exhaustive account of divine reality—whether the biblical God or the god of any other religion or philosophy. But it does represent a fundamental condition of both biblical and philosophical traditions of talk about God and is characteristic of so-called monotheism in both. At least in this area of tradition, all other statements about God rest on the assumption that this word refers to the all-determining reality. Then propositions about God can be tested by whether their content is truly determinative for all finite reality—as far as it is accessible to our experience. If this is the case, nothing real can be understood fully in its uniqueness without reference to the God thus proposed, and, in turn, one must expect that a deeper understanding of all reality is only possible in reference to the supposed divine reality. To the extent that both these things are true, we can speak of proving theological propositions.

# Is There Any Truth in God-Talk?

This casts some light on the differences between theological statements and unmediated personal statements of faith: Precisely because theological statements can be tested as propositions for their logical implications, that is, because of their scientific character, they—unlike faith statements—deal indirectly, not directly, with their real subject, God. On the other hand, they do thematize the element of knowledge in those unmediated statements of faith, which—as we have already mentioned—are themselves not simply performative utterances (they do not express a purely subjective commitment) but always imply a propositional element as well. Theological reflection thematizes the cognitive or propositional element contained in statements of faith; because it does so, it must, differently from unmediated piety, speak about God indirectly.

Of course, the question now arises how the totality of finite reality that is included in the idea of God can yield a practical criterion for testing theological statements. Where should the testing begin, and where should it end, if absolutely everything can be adduced as a criterion? And above all, what establishes the relative value of this or that detail for the whole of reality that is to be thought of as determined by God and to be tested in those terms?

Ancient philosophy originated in critical theology, which sought to understand the totality of all finite reality as cosmos in order to conclude from that to the basis of this totality, its *arche*. In our contemporary experience, on the other hand, reality as a whole is still unfinished. It exists in a process of becoming that is still incomplete. Hence, the sum of all that we know as reality—even if it could be comprehended by some human being or other—would still by no means be the genuine whole. The system of what now exists is, rather, an unreal whole (Theodor W. Adorno), if and because reality is not yet finished as a complete whole.

Nevertheless, people cannot live without the idea of an ultimate, all-encompassing whole because every individual thing we experience has its particular meaning only in the context of the whole to which it belongs. Similarly, every limited whole has meaning only as member of a larger whole. Hence, every experience of a particular individual thing already contains an awareness—usually unexpressed—of the all-encompassing totality of all reality. But the idea of an all-encompassing totality reaches beyond everything that is already present; its scope is not the totality of the cosmos as it exists, but rather it anticipates, as a totality of meaning, the not-yet-present completion to perfection of all reality and the power that makes such a perfection possible.

This merely anticipated (because not yet completed) totality of reality's meaning is at play, unexpressed, in every individual experience. Only in light of it do we experience the individual given thing as something determined in this or that way and as having a particular meaning. But the totality of meaning that illuminates all experience, even though the process of experience remains open and expansive, is expressed thematically in religious experience. Wherever people, as in religious experiences, have a religious encounter with the totality of their being, with the *universum*, as Friedrich Schleiermacher said, what is beheld in the finite content of that experience is found to be revealing. The totality of meaning of such experiences is made expressly thematic in religions—and in philosophies and worldviews that are in this respect related to religion—with regard to the powers that are determinative of all reality. Even the alterations of religious consciousness are founded in alterations in the experience of reality that reveal in a new form the total meaning of reality and thus the divine powers that determine all reality.

A testing of theological statements will, therefore, have to hold to religions and their changing history because the total meaning of reality against which the truth of theological statements must be tested is already thematic in religions. And the changes undergone by religions in the process of their history have steadily accomplished, in fact, what is to be accomplished expressly and methodically in the testing of theological statements: The propositions of religious tradition are measured by whether the reality of experience can be integrated by the gods of tradition, or whether other powers present themselves within the experience, powers before which one's own tradition collapses. Should the tradition prove unable to explain the experience, that then must be explained in a completely new way or through a modification of tradition or by adherence to a different tradition.

# Is There Any Truth in God-Talk?

## HOW CAN STATEMENTS ABOUT GOD BE TESTED?

A scientific study of religion[4] that would concentrate on such processes, by which, in the course of history, even religious propositions about the all-determining divine reality could be measured against experienced reality to see whether that determining power would show itself in that realm as well—such a scientific study of religion would examine the specifically religious in religions, the divine reality that proclaims itself in them and, at the same time, is a matter of dispute in the process of religious history. It would then be a *theology of religions* and not a mere psychology, sociology, or phenomenology of religions. In turn, theology itself is only possible in the context of such a theology of religions and their history because the totality of meaning of the real, against which theological statements must be tested, is explicitly thematic in religions and, because of the historicity and incompleteness of reality, is present only in the form of religious anticipation (in the broadest sense of the word *religious*). A critical examination of the history of religions as a history of changes in the experience of reality as a whole might then also ground the special character of one's own religious tradition in the context of the other religions. Likewise, one could examine the changes in one's own religious tradition—the history of Christianity—to see to what extent it has been affected by changes in the experience of reality and the anticipations of the totality of meaning of the real that are continually integrated in those changing experiences.

Through this process, the horizon of the problems connected with religious experience in each period can be made accessible, and theological propositions should be tested in relation to that problem-horizon. One would have to measure propositions about God as creator on the one hand by the experience of reality accumulated in the transmitted views of religious creation belief and, on the other hand, by the problems presented to creation belief by modern natural science and the worldview shaped by it—to the extent that the perspectives of natural science contain, unexpressed, an anticipation of the totality of meaning of reality.

We have seen that statements about God, and God's speaking or acting, cannot be directly tested by the measure of God's own reality, since God is not available as a measuring stick and, moreover, God's reality is a matter of dispute. Statements about God can be tested only against the implications of the *idea* of God, to the extent that God is posited as the all-determining reality. However, the total reality that is to be examined to see whether it proves to be determined by the divine power posited by theologians is not comprehensible to us as a completed whole, such as ancient philosophers envisioned when speaking of the cosmos. We can only anticipate the totality of meaning of experienced reality. It is true that, in every experience, such anticipation is in play and is even constitutive for the definitive content of every individual experience. But reality is only encountered thematically in religious experience, which has found its great and enduring forms in the historical religions, including the philosophies and worldviews that have arisen historically on the basis of religious traditions and in conversation with them.

The totality of reality against which statements about God would have to be tested is thus available to us only in the form of subjective anticipations of the unified meaning of all that is real, which, even unexpressed, determine all experience but which are expressly posited in the religions. Hence, religions constitute the immediate object of theology, which as a science of God is only possible indirectly, never directly. We should test religious traditions because in them we find expressly thematized what in all other experience is only implied. Theology, therefore, hypothesizes about the extent to which the statements of religious tradition, from the origin of those traditions to the present, make explicit what can be shown as implicit anticipation of meaning in other experiences.

Such initiatives contain the beginnings of a theological methodology that cannot be described in detail here. But this much should be clear: Such a theology does not presume any specifically Christian faith convictions as its "foundational context" (as termed by Gerhard Sauter), no matter how much the "context of discovery," which motivates individual theologians in constructing their hypotheses, may be shaped by Christian tradition. The example of Karl Barth has shown that confusing the context of discovery and the foundational context destroys the propositional meaning of theological statements. Within a theology's foundational context, as sketched here, the unique character of Christianity and its truth consciousness must first be treated as problematic, no matter how much the theologian may be subjectively persuaded that the superiority and truth of the Christian faith will be maintained by the process of theological investigation, both in the world of religions and in experience as a whole.

Thus, we leave the ground of confessional theologies, in which everything is predetermined by the particular faith position of that confession and which is not open to public debate. Today, a theology that only rationalizes prior judgments can no longer claim intellectual seriousness and scientific legitimacy. Theology can be taken seriously only if, within its foundational context, it accepts that its assertions must be tested against reality. This occurs when we understand its statements as hypotheses to be tested in an appropriate way and that their truth cannot be presumed in advance. By means of such a modest self-assessment, theology would acquire not only legitimacy in terms of philosophy of science but would open up and expand its own thematic field, without losing anything of its subject. Such a self-critical assessment of its approach is appropriate both to the pluralism of modern society and its public educational institutions and to the demands of a fair dialogue with other religions.

In conclusion, I offer one remark regarding the question of how theology can speak of testing or evaluating its hypotheses. A strict verification in the sense of logical positivism, by tracing theological statements to sense observations, is certainly impossible. But such a strict verification is not possible even for the posited laws of physics because no general rule can be exhaustively tested by a finite number of cases to which it applies. Propositions regarding laws, on the other hand, can be falsified by being breached in a single demonstrated case: Their claim to absolute universal validity can be refuted by a single exception. But even this principle of falsification, developed by Karl Popper, cannot be applied to theology (or to other disciplines, such as history) because its characteristic propositions are not general propositions about universal laws.

However, it is not clear why the concept of science must be restricted to the knowledge of general rules. Why should it not include the knowledge of what is individual and unique? It is equally unclear by what right the concept of truth and the related notion of verification should be reduced to sense observations when they are more properly applied to the whole of reality and of experience. We would do better to speak of verification in the broader sense of an evaluation of hypotheses by testing them against all relevant circumstances.

The broader concept of verification has also been adopted in theology, although the suggestions thus far presented are not fully adequate. When John Hick speaks of an eschatological verification in which the promises of faith will be revealed as true at the end of the ages, when the reign of God comes in glory, he is right in one sense; but that is of little use to anyone who wishes to make a judgment, even a provisional one, about the truth content of religious propositions in the present. On the other side, Gerhard Ebeling has offered an original interpretation of verification by suggesting that God verifies human beings by bringing them into truth. This argument, too, touches a criterion of the truth of faith, in which the truth of God is proven, but it does not overcome the barriers of faith's subjectivity. The possibility and the testing of theological propositions are not made one whit clearer by reference to such existential verification. Certainly an ultimate proof of theological propositions is unattainable—in contrast to faith's trusting certainty, which is of a different order. But a preliminary evaluation of theological hypotheses is certainly possible and may shed light on the problems and themes of religious traditions and on the implied meanings of present experience.

PART TWO

# CREATION AND NATURE'S HISTORICITY

# ⊹3⊰

# The Theology of Creation and
# the Natural Sciences

HALF A CENTURY AGO, Karl Barth wrote, in the foreword to his theology of creation, that ". . . there can be no scientific problems, questions, objections or aids in relation to what Holy Scripture and the Christian Church understand by the divine work of creation."[1] The price of restricting creation theology to the "repetition" of biblical statements on the subject was that it could no longer be clear to what extent the biblical creation faith applies to this world in which today's humanity lives, and to the world described by the modern natural sciences. The proposition that the world was created by the God of the Bible then becomes an empty formula, and the biblical God himself a powerless specter, since he can no longer be understood to be the origin and perfecter of the world as it is given in our experience. We should agree rather with Karl Heim, as Heim attempts to relate theological statements about the creation of the world and the world's end to the corresponding present-day ideas of natural science.[2] In English theology, as early as 1889, the collection *Lux Mundi*, edited by Charles Gore,[3] began to apply Darwin's theory of evolution to theology by associating evolution with the biblical notion of salvation history culminating in the Incarnation, and this point of view has maintained its influence to the present, together with associated impulses originating in the work of Pierre Teilhard de Chardin.

## CONSONANCE BETWEEN CREATION THEOLOGY
## AND NATURAL SCIENCE

Despite all the associated difficulties, Christian theology cannot avoid the task not only of asserting that this world, the object of natural science's descriptive work, is God's creation but also of making it understandable as such. This fact does not mean that theology should or may intervene directly in the discussions of natural scientists at the level of scientific descriptions and theoretical constructs. Theological interpretation of the natural world as creation cannot present itself as in competition with physics or any other natural science. That is excluded by the very fact that theological argumentation proceeds on a different methodological level than hypotheses regarding the laws of natural science and their experimental verification. From a theological perspective, a world-event appears as a unique and irreversible history that, as such, is the expression of divine action even though, in the course of that history, uniformities and orders of events may correspond to the notion of the laws of natural science. Thus, in the book of Genesis, after the story of the flood, we read: "As long as the earth endures, seedtime and harvest, cold and heat, summer and winter, day and night, shall not cease" (Gen. 8:22). But such regularities in events are regarded as the product of a unique divine decision, not as the expression of a timeless ordering of natural events according to natural laws. The theological orientation to the historically unique and to the unique and irreversible process of history is also connected to the fact that theology does not see space and time as a uniform arrangement of spatial and temporal units that, as such, is geometrically conceivable, numerable, and measurable. The mathematical form of conception of natural process and the natural-scientific concept of laws are linked. That theology does not use mathematical models, in contrast, does not only reflect theologians' incompetence in that regard, but also the fact that a different perspective is appropriate to their discipline.

Is this an example of qualitative assessment, something that, in the history of modern natural science, has repeatedly been traced back to quantitative (and thus also mathematical) descriptions? The ideas in the biblical account of Creation concerning the sequential appearance of natural forms have, in fact, been replaced in modern natural science by concepts founded in a quantitative description of natural processes following certain laws. Does this have some fundamental significance for the overall relationship between theology and natural science? That theology as a whole must ultimately be replaced by physics is something that has repeatedly been asserted by an American physicist, Frank Tipler. In his book *The Physics of Immortality*, Tipler attempts to show that the history of the universe is moving toward an omega point characterized by central features of the traditional idea of God and should be understood not only as an end point but also as a creative starting point of the movement of the universe and the locus of an identical replication of all forms of intelligent life in the dimension of eternity.[4] He bases these propositions in a mathematical theory of physical cosmology. For laypersons, certainly, the multiplicity and variety of models in scientific cosmology in the last several decades are impressive. It appears to be a highly speculative discipline. But what should be theology's attitude to the possibility of such forms of argumentation?

Attempts to transform theology into physics should be regarded both with interest and with a certain skepticism: With interest and openness because such attempts do, in their own way, serve to counteract the widespread notion that theological and physical ideas exist alongside one another without any mutual relationship—a prejudice that usually results in theology's being judged irrelevant for understanding the reality in which we live. But one should also approach such attempts with skepticism because of the apparent incommensurability of natural science's knowledge of laws and the theological approach to the world. Should the concept of the world as a unique and irreversible history of constantly new and irreducible events, including the divine act of the world's origination and Christians' eschatological hope for the future, be completely dissolved into a description based on the concepts of natural science? I consider theological anxieties about this scenario to be unfounded. At any rate, there is a historical parallel to such a concept of the world in Aristotle's physics, which, however, posited as part of its subject field the existence of God but not a future resurrection of the dead. Certainly, an inappropriate notion of God as the origin of the natural universe would be one that did not regard God so much as the exponent of cosmic processes, but rather sought to understand the creation of the world in terms of its origins in God. Christian theology, however, would be expected to view such knowledge of creation, which embraces all aspects of created reality, in connection with the vision of God enjoyed by the blessed at the end of time. Temporarily, human perceptions of the world are conditioned by the finite nature of our knowledge and are thus available only through conjectures and through the testing and revision of those conjectures. Christian theology, on the other hand, attempts to think of God as Creator of the world on the basis of the revelation of God in Jesus Christ—but in doing so, it is far from being able to explain worldly events in all their details.

Thus, a rapprochement between creation theology and natural science should probably be described more as a consonance of the two ways of perceiving the world than as an attempt to base one conception on the other. Consonance presupposes an absence of contradiction, but it demands more than that. Absence of contradiction can also exist in the case of ideas that exist alongside one another without any mutual relationship. Consonance, on the other hand, includes the idea of harmony, that is, a positive relationship. How can such consonance be claimed for propositions developed on quite different methodological levels, especially when those methodological levels cannot be directly linked to each other? In such a case, it is necessary to seek yet a third level to which both of the others are related. Such a third level for the dialogue between natural science and theology has, in fact, always existed, namely, in philosophy.

When natural scientists speak of the relevance of their findings and theoretical formulae for our understanding of reality, they are already in the realm of philosophical reflection on the processes and results of their sciences and no longer, in the strict sense, on the level of scientific argumentation. Such reflections on the relationship between natural law and the contingency of events, on causality and freedom, on matter and energy, on the concepts of time and space, or on development inevitably are couched in the medium of philosophical language and its history. Moreover, the fundamental concepts of natural science are derived, as a rule, from the language of philosophy, modified to meet the needs of scientific usage. Studies of the history of basic scientific concepts such as space, time, mass, force, and field have made the connections between the philosophical meaning of these terms and their scientific use clear. A knowledge of the history of science, especially the history of the terminology of the natural sciences, therefore belongs—together with an overview of philosophical discussion of these themes—to the preconditions for a fruitful dialogue between theology and the natural sciences.

Theology in turn, at least Christian theology, has had a close relationship to philosophy throughout its history, though that relationship has not been free of complications or tension. Unlike the case of the natural sciences, this relationship does not first and foremost have to do with the philosophical origins of theological terminology but with the integration of philosophical discourse about God, the world, and humanity, a discourse that needs to be completed in theology's explication of the revelation of God as the creator and perfecter of the world and human beings. Such an integration of philosophical statements and questions in Christian theology has always been associated with a more or less drastic transformation, resulting, in the course of history, in tensions between theology and philosophy. Still, theology has always depended on philosophy for its assertion of the truth of the biblical God and his revelation, binding on all people, namely, by adopting—though critically—not only philosophy's teachings about God but also its assertions about the world and humanity. At this point, it is also clear how theology's relationship to philosophy—that is, to philosophy's interpretation of the world—constitutes the basis for Christianity's dialogue with the natural sciences: The integration of scientific techniques of observation and reflection on the overall conception of the reality of the world and the human situation are not only themes of theological doctrines of creation but have always been an aspect of philosophy's approach to the world and the development of a philosophical world-concept. When theology engages in a critical adoption and adaptation of a philosophical view of the world, it is simultaneously concerned with the awareness of nature that is integrated therein, and the theological transformation of philosophical concepts of the world must, just as for philosophy, be measured by its capacity for correctly incorporating the methods of observation and results of the natural sciences.

Unfortunately, at the present time, most philosophers neglect their obligation to engage in natural philosophy as a summary reflection on the scientific description of the phenomena of nature. The resulting gap is being filled by natural scientists who attempt to offer a philosophically reflective orientation to the reality of the world from the perspectives of their several specialties. However, it often happens that the philosophical problem-horizon of the respective themes, along with the history of the problem in philosophy, is not adequately considered. It is then a task of theology, in dialogue with natural science, to recall the philosophical problem-horizon of the themes in question and, within that framework, to bring to bear the specifically theological accent on these themes.

## THEMES FOR A DIALOGUE BETWEEN CREATION THEOLOGY AND THE NATURAL SCIENCES

In the remaining parts of this essay, I will give examples of what I have been saying thus far in general terms about the dialogue between theology and the natural sciences, drawn from topics that seem to me especially important for this dialogue because they are fundamental to our understanding of the world itself. First, I will briefly examine the concept of law in relationship to the contrary idea of contingent events. The mutual relationship of these two aspects in the description of natural events can be shown through the concept of natural law itself, but it also opens the possibility for Christian theology to apply the specifically biblical conception of reality to the description of natural reality through natural laws. Second, I will discuss the concepts of space and time, which are basic not only to scientific descriptions of nature but also to theological statements about the relationship between God and the world. This is followed, third, by the question of the relationship of God to the movements of bodies, their origins and decay: the classic theme of scientific descriptions of nature within the framework of the concepts of space and time. A clarification of the relationship of God to space and time, therefore, can be expected to have consequences also for our understanding of the existence and movement of created reality in its relationship to God within space and time. Finally, in this connection, we will also address some ways of viewing the relationship between creation theology and evolution, not only in regard to the evolution of life but also its situation within the history of the universe.

## Natural Law and Contingency

In 1970, I published an essay on the topic of "Natural Law and Contingency" that for the rest of the year was the object of intensive discussion by a group of physicists and theologians; it received significant modifications as a result of those conversations.[5] Theological interest in the topic was based on the fact that the biblical accounts of God's action in history emphasize what is new and unexpected in divine action—a characteristic that applies to God's creative action as well. The history of divine action is a unique and irreversible sequence of such contingent acts. In turn, the concept of contingency thus applied to divine action in history is philosophical in origin, a description of the accidental or possible in contrast to the necessary. But while for Aristotle the concept of contingency was associated with matter, medieval Christian Aristotelians, especially Duns Scotus, had applied it to God's freedom to will and to act. The concept of natural law applies first of all, logically, to conditions of its application that are contingent in relation to it, that is, the initial and boundary conditions of the process described by the law. The initial and boundary conditions of application of a formula of law may be the result of constellations of things that are again described by natural law, but that does not alter the fact that every such description again presupposes contingent conditions of its application, so that it seems proper to understand natural laws as descriptions of uniform structures of processes appearing within what is contingently given. The supposition this implies, that *all* that happens is primarily contingent, even when uniformities appear in the sequence of events, was seen by natural scientists in the 1960s as problematic, even though such a notion was suggested also by the irreversibility of time. Since that time, the contingency of events that was then being discussed (as distinct from contingency in the purely logical sense) has been generally acknowledged in light of so many chaotically developing natural processes, and in view of quantum physics' concept of indeterminacy, contingency can be regarded as the fundamental character of every elementary event, if we grant that the same events, because of the uniformities in their sequence, can also be the objects of natural-law descriptions. In turn, the possibility of description in terms of natural laws does not eliminate the fundamental contingency of individual events; rather, the observable uniformities in the sequence of events, describable in terms of hypothetical laws, appear in themselves as contingent facts. But while theological statements about created reality and God's actions in creation refer primarily to this contingent aspect of events, the natural sciences' description is interested in demonstrating the conformity of the processes to laws, although the pertinence to contingent facts is itself constitutive for the applicability of the concept of law.

Those who participated in the conversations in Heidelberg in the 1960s seemed thus to have found a common basis for the conversation between theology and the natural sciences, beyond vague analogies and metaphorical transfer of concepts from one discipline to another. However, this agreement about natural law and contingency did not lead to more specific understandings of natural reality from a theological perspective. For that we need to acquire some kind of theological access to the basic physical concepts of force and motion, as well as to the underlying concepts of space and time.

## Space and Time

In the early eighteenth century, there was a dispute among natural philosophers about the concept of space; theological implications played a crucial role in that debate. The correspondence between Gottfried Wilhelm Leibniz and Samuel Clarke over Isaac Newton's description of space as *sensorium Dei* in his *Optics*, which appeared in 1706, is still of more than mere historical interest today. Even though Albert Einstein's theory of relativity has made Newton's concept of absolute space obsolete, Newton's thoughts about space and God's relationship to space were very complex. It is worthwhile taking a closer look at what in his thinking has been rendered obsolete and what has not. The notion of absolute directions and relationships between quantities within space, independent of the masses moving within space, is undoubtedly out of date. But in Newton's and Clarke's conception of the relationship between God and space, there is an idea that is still relevant today. Clarke defended Newton's association of the concept of space with the idea of God against Leibniz's objection that, in that case, God must be divisible and composed of parts. Clarke's main argument was that infinite, undivided space precedes all divisibility in space, and it is the former—not geometric space, which is composed of parts—that is identical with divine immeasurability, by means of which God can be present to each of his creatures in its particular place. This argument was taken up by Immanuel Kant in his *Critique of Pure Reason* (1781): The view of space as an infinite whole is presupposed by all notions of particular spaces (A 24–25). Kant did not pursue the theological implications of this idea because he regarded space as a purely subjective form of human conception. But anyone who considers this subjectivism questionable will immediately be confronted—like Samuel Alexander in the twentieth century—by the theological implications of the priority of an infinite, undivided space over all concepts of finite space. The point is that the infinite space presupposed by every division of space is itself undivided, in contrast to all geometric concepts of space, which are constructed on the basis of units of measure: Every geometrical unit of measure is itself a portion of space that, as such, presupposes the undivided whole of infinite space. The latter, however, is an infinity that cannot be conceived, as in geometry, as the unlimited addition of units of measure but one that precedes all division, and thus all measurement. Benedictus de Spinoza's mistake in conceiving space as an attribute of divine substance was that he did not distinguish infinite geometric space from the infinite and undivided space of divine immeasurability that is presupposed by all geometry. If this distinction is kept in view, there are no pantheistic consequences such as those Leibniz appears to have suspected in Newton. The transition to the space that is distinguished by parts and places then occurs with the appearance of finite entities in space and their relationship to one another. In this way, the relativity of spatial relationships can take into account masses moving within space. But every space that is divided and consists of parts always, as Kant emphasizes, presumes an undivided spatial whole because parts and division are only possible within a given space, which thus also precedes every geometric concept of space. Notions of divine immeasurability and omnipresence to creatures can be applied to this undivided space, as was the case for Newton and Clarke, without impairment of divine transcendence in respect to the world, contrary to Spinoza's conception, to which, as we know, Einstein felt himself bound, but that made no distinction between the undivided, infinite space of divine omnipresence and geometric space.

Analogous to the relationship between space and God's immeasurability is that between God's eternity and time. In his transcendental aesthetics, Kant treated time very much as he did space; the endless and undivided entity of time is the condition for all divisions of time and every idea of such divisions: "Different times are only parts of one and the same time" (A 31). The undivided whole of time, or rather the whole of life, which appears as separate moments in the sequence of time, has been called eternity in philosophical and theological tradition since Plotinus' treatise on time. Eternity, said Plotinus, is the "undivided perfection" (*Enneads* II, 7, 11) of what appears as divided in the sequence of time. Boethius, who handed on this idea to those who came later, described eternity as the simultaneous and perfect presence of life without limits (*interminabilis vitae tota simul et perfecta possessio* [*De cons. Phil.* V, 6, 4]). Eternity is thus not timelessness in the sense that eternity and time are altogether alien to one another. Rather, even according to Plotinus, time is grounded in eternity because the transition from one moment of time to another can be understood only in terms of the whole of what is divided into time, a whole, that is, eternity, which is still present in the separation and sequence of moments of time. The same thought is present in Kant's statement that different times are only parts of one and the same time. However, Kant no longer grounded time in the unity of eternity, but, analogously to space, in human subjectivity, and in particular the "standing and lasting" human "I" (A 123) as the basis for the unity of all experience. But in view of the temporality of the "I" itself, of which we are aware in our self-consciousness, Kant's grounding of time in the unity of the "I" seems to be substantially more problematic than Plotinus' grounding of time in the concept of eternity.

For a theological view of nature, then, God's eternity is present in time, as origin and perfection of time and all things temporal—he is the origin in the sense of the condition for the continuity of what is divided in the sequence of time and the perfection because everything temporal seeks from the future the realization of its wholeness: Through the future, eternity enters into time.

For time as for space, then, we see that these concepts cannot be established in terms of the question of measurement by clocks and spatial units. This should be an important point in a conversation between theology and the natural sciences because natural science's interest in time and spatial relationships is connected to the possibility of measurement. But the concepts of space and time are more fundamental than efforts at measuring them. If that condition is ignored, we run into contradictions. All units of measurement are themselves partial times and partial spaces that must have temporal and spatial limits and, therefore, already presume time and space.

## God's Activity in Nature's Activity

Much more difficult to address than the question of the relationship of space and time to divine immeasurability and eternity is that of the relationship of God to the forces at work in natural events. Yet it is a crucial question for a theology of creation grounded in the Bible because this is the question of the possibility of God's acting within his creation, not only at the beginning but in the whole course of its history. This is the point over which Christian theological descriptions of the world and those of the natural sciences became estranged in the seventeenth and eighteenth centuries. The crucial issue was that a mechanistic interpretation of natural processes, begun by René Descartes and accomplished, contrary to Newton's intentions, in the course of the eighteenth century, traced all evidences of force to bodies and their influences on one another. This conception excluded God, in principle, from any interpretation of natural events. For if there was one point on which modern philosophical theology was in agreement with the Scholastic doctrine of God, it was that God cannot be a body. Therefore, the idea of effects of force emanating from God, and therefore of divine action in natural events, was excluded from the outset. God was thus cordially excused from the natural world.

Only when one considers the wide-ranging consequences for an atheistic picture of natural events of tying all moving forces to bodies and masses can one also measure the potential implications of introducing concepts of field in the description of natural processes, and this has been the case since the discoveries of Michael Faraday. This does not mean that the demonstration of electrical and magnetic fields was immediately useful as a model for understanding God's working within nature. But although the influence of fields has, to a large extent, a correlative in masses, since Faraday's time the goal has been to interpret all physical phenomena as manifestations of fields. Such ideas were congruent with Newton's intention to show that moving forces are ultimately not material in nature and do not emanate from bodies. Thus, Newton also thought of God's action in the universe by analogy with the way in which our spirit moves the members of our body.

However, an application of the concept of field in theology is not suggested primarily by the question of God's action in nature, but by the internal problems of the doctrine of God. The designation of the divine being as "Spirit" in the Gospel of John (John 4:24) has, of course, been interpreted since Origen to mean that God is *Nus*, that is, a rational being existing without a body, but this platonizing interpretation does not match the biblical word *pneuma* or the Hebrew word *ruah* which lies behind it. Both of these have the basic meaning of air, breath, or wind. In Greek thought, the word *pneuma*, which we translate as "spirit," was used both in pre-Socratic philosophy by Anaximenes and by the Stoics in the sense of exhalation. In Stoic doctrine, the air, which, as the finest of all matter, penetrates everything, through its own "tension" (*tónos*) holds the whole cosmos together. Likewise, early Christian theologians before Origen still understood the New Testament description of God as in this sense. Now, one of the best-known scientists of our time, Max Jammer, who studied the history of a number of basic concepts in physics, regarded the doctrine of *pneuma* in antiquity as the predecessor of the concept of field in modern physics. In fact, a force field, in its visible sense, is first of all air filled with states of tension. However, modern concepts of field differ in a crucial respect from antiquity's ideas of *pneuma*; the influences of field do not require—as was still presumed at first even in the nineteenth century—a material medium such as the air or an "ether." They can expand in space without such a medium. However, the materialism of the Stoic doctrine of *pneuma* as air, even in the sense of the finest, all-penetrating matter, constituted for Origen the basis for rejecting that idea as an interpretation of the Johannine description of God as Spirit. The absurdities of conceiving of God as a body—and thus as divisible and composed of parts—constituted the ground for interpreting *pneuma* as *Nus*, that is, for regarding God as a bodiless rational being. But it is clear that this idea does not comport with the sense of the word *pneuma*. At this point, the concept of field, as a substitute for the ancient doctrine of *pneuma*, is helpful for theology because it permits us to separate the sense of the word *pneuma* from the idea of a material substratum or medium. If the divine being is understood as a field that manifests itself in the three "persons" of the Father, the Son, and the Holy Spirit, we can do justice to Origen's objections to an idea of God as body and still hold fast to the sense of the word *pneuma*.

Is such a theological use of the concept of field merely a matter of metaphor? At first glance it may seem so. Nevertheless, one fundamental requirement for the applicability of the idea of field is fulfilled by theology, namely, a relationship to space and time, although in this case in the sense of discourse about the undivided, infinite space of divine immeasurability, presupposed by all geometric descriptions of space, and of the undivided unity of time in divine eternity as a condition for the possibility of every temporal consequence. The interpretation of the pneumatic essence of God's divinity as a field can be applied to the undivided unity of space and time that precedes all geometric description. It is thereby at the same time distinguished from physics' concepts of field but should be regarded as the precondition for them, analogously to the situation with regard to space and time. The field of divine omnipotence thus does not compete with the field-entities of physics but works through and beyond the forces of nature without being exhaustively expressed by them. As divine omnipresence is not tied to the speed of light, so the field effects of divine omnipotence require no mediation through waves. The concept of expansion in the form of waves, so important for the quantitative description of field effects in physics, at least for the concepts of field in classical physics, is not constitutive of the concept of field as such, whereas the concept of a field without relationship to space and time would lack any content at all. If we can conceive of the concept of field without a reproduction of field effects through waves, we can also conceive of a nonlocal, instantaneous communication between phenomena as a field effect.

### Creation and Evolution

It is not possible within the compass of this essay to apply what has been said to a theological interpretation of the world of creatures as they appear in the history of the universe. I have presented a proposal for such an interpretation in my treatment of the doctrine of creation within my systematic theology. But it is more and more important today for the dialogue between theology and the natural sciences that they come to an understanding about the bases for such interpretations. Here let me just say this: The key to the connection between eternity and time lies, as I have suggested, in the meaning of the future for our understanding of what exists in time. Through the future, eternity enters into time. From the future continually emerge new, contingent events, and, in turn, everything that exists can only await and receive the possible entirety of its being from the future. All things advance toward the reign of God, but God's rule is already at work, out of God's future, in the present of his creatures. From the point of view of creatures, this relationship is reversed. The future becomes a field for the extrapolation of what is present and what is known from the past. The same is true of the history of the universe. A mythical worldview sees the order of the universe as grounded in its beginnings. This is evident also in the biblical account of Creation, although its literary form is no longer that of myth. But the image of all created things originating in an initial seven-day week is in tension with the perspective that is otherwise characteristic of biblical thinking: that God is acting ever anew within history toward a future perfection of the creation. The notion of an order of creation that was finished at the beginning and not altered thereafter has long put a strain on mutual agreement between theologians and natural scientists, especially in the period of struggle over the theory of evolution.

# Theology of Creation and the Natural Sciences

But much more important for a consonance between creation theology and natural science is that the evolution of life represents an irreversible process in which new, contingent events are continually occurring. Something similar holds for the history of the universe. Both with regard to the origins and evolution of life and in the realm of cosmology, the barriers between the natural sciences' understanding of the world and that of Christian theology raised by a particular worldview have fallen. It would be asking too much of scientific cosmology to expect it to produce a proof of the existence of God such as Pope Pius XII, in his first enthusiasm over the now-standard model of an expanding universe, thought he perceived therein. It suffices that the theological interpretation of the worldly events as creation can be developed in consonance with actual developments within the natural sciences. For this it is necessary that the theological doctrine of Creation remains open to new learning, not in the sense of an external, apologetic accommodation to changes in the scientific picture of the universe but in the sense that theology continually develops interpretations out of its own resources that seek to be true to the changing knowledge of our experience of the world, so as to integrate that knowledge into the Christian understanding of the world as the creation of the God of the Bible.

# ⇥4⇤

# Problems between Science and Theology in the Course of Their Modern History

THERE WAS A TIME, around the turn of the nineteenth century, when many talked about a "warfare" of science with Christian theology, the title of a book by Andrew White, first published in 1896.[1] It was a time when, under the lead of classical physics, modern natural science felt victorious. The warfare was said to have started with Copernicus and Galileo. But while there certainly had been a tension between the heliocentric worldview and the authority of the Bible, the term *warfare* is much too strong. Neither Copernicus nor Galileo intended such a conflict. The proceedings against Galileo at Rome and their unfortunate results did a lot of damage but did not undermine the positive attitude of scientists to the Christian faith. Besides, the proceedings were due more to the issue of church authority than to the scientific position of Galileo. On the Protestant side, Martin Luther, in one of his table talks, said in 1539 that he would rather believe Holy Scripture that reports in the book of Joshua (10, 12–13) that Joshua ordered the sun to stand still and not the earth, which presupposed that, according to the order of nature, the sun would go around the earth, rather than the reverse. Biblical literalism continued for a long time to prevent Protestant theologians from adopting the Copernican view, although Johannes Kepler as early as 1596 suggested that the point of the story of Joshua might simply have been that he prayed daylight to last long enough, until the battle against the Amorites was finished. There was no real conflict, then, between the book of nature and the book of Joshua.

The more serious problems between science and Christian theology emerged somewhat later, with the introduction of the principle of *inertia* in the natural philosophy of René Descartes. The key importance of that principle is to be perceived in the context of Descartes' concept of motion, which was closely related to his conception of bodily reality. He considered movements (like rest) a property of the bodies that are in movement (*Principia* II, 31–32). God in the beginning created matter together with movement, and he endowed every creature with its movement. Because of his immutability, then, God preserves the amount of movement and rest as it was created in the beginning (II, 36), without further interfering in the interactions of his creatures. Extrapolating ideas of Galileo before him, Descartes disregarded the concept of God as *final* cause of the universe and relegated his relationship to the world "to the position of first cause of motion, the happenings of the universe then continuing in aeternum as incidents in the regular revolutions of a great mathematical machine," as Edwin A. Burtt described Descartes' natural philosophy in 1924.[2] Burtt's judgment is correct, that the resulting picture is "fundamentally different from the Platonic–Aristotelian–Christian view" of the world that had been "centrally a teleological . . . conception of the processes of nature."[3] But the decisive point in this opposition was not the abolition of final causes and the corresponding emphasis on God as only efficient cause of the universe. Rather, the decisive change was that Descartes denied any further interference of the Creator with his creation under the pretext that such was required by divine immutability. Descartes did not deny that the creation is in continuous need of conservation by its Creator, but because of his immutability, God preserves the creation precisely in the state in which it had been conceived in the beginning. Therefore, God is not the cause of changes in the world of his creation. All changes result from the mechanical interactions of the creatures among themselves, when they transfer their movements upon one another.

The first principle of this mechanistic view of the processes of nature and of its development was the principle of *inertia*, the first law of nature, as Descartes said (II, 37). It affirms that everything tends to persevere in its state, be it a state of rest or of movement. This principle changed fundamentally the Aristotelian-Scholastic theory of movement, which affirmed that all movement tends to rest. This traditional view required an extrinsic cause for any movement: *Omne quod movetur, ab alio movetur,* said Thomas Aquinas (*Summa theologica* 1, 2, 3, respectively). In the view of Descartes, such an extrinsic cause of movement was no longer necessary in general, since movement now was conceived as a "state" of the body as rest was. The concept of state came to function as the general term, to which movement and rest were subordinated. Accordingly, God's activity of conserving his creatures was understood to preserve them in the state of movement they had received in the beginning, in the act of creation. The consequence was, I repeat, that because of God's immutability all changes in the world of nature were conceived to be due to the mutual interactions of the creatures, not to any divine intervention. Thus, the principle of *inertia* made the world of nature independent from all further divine activity. The consequence was deism, with a God who only acted as creator in the beginning, while the further course of nature was left to its own mechanisms.

It was this view of the world that caused Isaac Newton's suspicion of atheistic consequences of the Cartesian worldview.[4] As a remedy, he introduced his concept of force as cause of movement. In his *Mathematical Principles of Natural Philosophy* (1687) Newton reformulated the principle of *inertia* in terms of a force that is intrinsic in bodies, *vis insita*; he distinguished this from forces that act upon bodies from the outside, *vis impressa*; and among these he admitted other than mechanical forces. From the preface of his *Mathematical Principles*, one could infer that his ideal was a completely mechanical description of nature. But in his *Opticks* (1706), Newton emphasized that the first cause is not mechanical and that there are other nonmaterial forces like gravity.[5] Thus, Edwin Burtt wondered "how was it that Newton historically came to pose as the champion of the more rigid mechanical view of the physical realm."[6] Contrary to Descartes, Newton affirmed in his famous *General Scholium*, which was added to the second edition of the *Principia* in 1713 (II, 31ff), that God not only created but also "governs all things" and that in him all things are "moved." In his *Opticks*, Newton affirmed that God moves everything as we move "the parts of our bodies" by our will. Burtt summarized Newton's position: God "is the ultimate originator of motion and is able at any time now to add motion to bodies."[7]

The development of physics in the eighteenth century did not follow in the line of Newton's combination of his natural philosophy with religious ideas. When Immanuel Kant published his ideas on a purely mechanical origin of the solar system, he contributed to a tendency that rendered superfluous the assumption of a creator God as efficient cause of the universe, as became evident in the completion of Kant's theory by Pierre Simon Laplace with his famous dictum that he no longer needed the hypothesis of a God as creator of the universe. When, as a matter of principle, the concept of force was attributed to bodies' exercising forces onto one another,[8] a return to the Cartesian mechanistic worldview took place, which Newton had intended to correct because of its implicit potential of atheism. In the eighteenth century, this tendency became victorious. The principle of *inertia* no longer needed a basis in the assumption of divine immutability. Newton himself made it independent from such an assumption when he defined the tendency of bodies to continue in their "state" of either movement or rest in terms of a *vis insita*. Soon the concept of force in this description was replaced by characterizing inertia as simply a property of bodies. The tendency of this principle to emancipate the description of nature from any need for divine intervention or succor was now complemented by the attribution of forces to bodies exclusively. This excluded by definition any divine intervention in the course of nature, since, whatever God may be conceived to be, he is certainly not a body.

The mechanistic interpretation of the course of nature, then, contributed decisively to the alienation between theology and the natural sciences. As a corollary it may be mentioned that David Hume's famous rejection of miracles in his *Inquiry Concerning Human Understanding* (1748, chap. 10) presupposed a view of nature where divine interventions were excluded in principle, as was the case in the mechanistic natural philosophy of his time. Hume was correct to criticize a conception of miracle in terms of a "violation of the laws of nature," but rather than replacing this misleading idea by the Augustinian notion of miracle as an extremely unusual event, he rejected the notion altogether as "most contrary to custom and experience," which was a far cry from John Locke's attitude concerning this issue as well as from Newton's.

Considering the importance of the attribution of forces to bodies in this conflict with theology, one might expect that Michael Faraday's introduction of a field concept of force could have reversed the picture. Faraday's tendency to envisage bodily phenomena as subordinate manifestations of fields[9] could have brought about a change in the relationship of science to theology. While God could not be imagined as body to influence the course of natural events, the divine Spirit could very well have been conceived as field, especially in view of the fact that the ancient Stoic ideas about *pneuma* as a cosmic principle, which, according to Max Jammer, were the "immediate" conceptual precursor of the field concept of modern physics,[10] were remarkably close to the root meaning of the biblical concept of spirit as "air in movement" like breath or wind. But in the field theory of James Clerk Maxwell, fields were conceived as correlate to bodies or masses; and when Albert Einstein renewed Faraday's idea of a priority of fields with regard to bodily phenomena, he did so by way of a geometrical description of the cosmic field of space-time, no longer in the sense of Faraday's conception of a field of force.[11] Thus, the potential of the field concept for the dialogue between science and theology has not been used until very recently by some theologians (for example, by Thomas Torrance and myself).

In the heyday of a mechanistic description of the universe of nature, the full weight of rational argument for the existence of God fell upon the teleological argument. Indications of design or teleonomic order, especially in the world of organisms, could be taken as hints or evidence of the existence of a designing mind, who created them. This is the background of the otherwise astounding passion in the debate about Charles Darwin and his theory of evolution, when he "showed that adaptation can be explained by random variation and natural selection."[12] Although there were positive responses by theologians to Darwin's theory very early, like the volume *Lux Mundi,* edited in 1889 by Charles Gore, in some circles, the hostile reactions of theologians to the idea of a natural evolution of organisms continues to the present day, although the biblical creation story itself says that God created vegetation and even the animals by commanding *the earth* to bring forth such creatures (Gen. 1, 11, and 24). In his act of creation, God did not need to do it all by himself, alone; he recruits the assistance of his creatures. In affirming that the earth brought forth not only primitive forms of life but also the higher animals, the Bible is more audacious than Darwin was. It is hard to see, then, what biblical reasons theologians can have to object to an origin of organic life from inorganic matter, not to speak of the further evolution of higher species from lower ones. Certainly, the biblical record did not yet employ an idea of evolution, but rather affirmed the constancy of species as a consequence of the conception that the order of creation was intended to remain as it was founded in the beginning. Nevertheless, the use of "the earth" as agency in God's work of creation is significant, especially when connected with the idea of a continuing divine activity of creation, as suggested by other biblical passages.

To critics of the Darwinian theory of natural selection, that doctrine could appear at first as another triumph of the mechanistic description of natural processes. This explains the enthusiastic response to the Darwinian theory among materialist scientists like Ernst Haeckel in Germany. The further development of the concept of evolution in terms of "emergent evolution" and "organic evolution" clarified, however, that the theory did not advocate a biological determinism but allowed the emergence of novelty in the course of evolution. Yet as late as 1970, Jacques Monod, in his book, *Chance and Necessity,* did not appreciate the positive value of chance and contingency for a theological interpretation of the process of evolution.[13] Chance was only called upon in the service of destroying the argument for design. But in a theological interpretation of nature, the element of chance or contingency is even more important than design because contingency and the emergence of novelty correspond to the biblical view of God's continuously creative activity in the course of history and in the world of nature.

Contingency and novelty in natural processes can be interpreted theologically as evidence of God's continuing creative activity. Just as the first act of creation expresses the freedom of the Creator, so does his continuous creative activity, which is manifested by the element of contingency in natural processes. For this reason, I emphasized the issue of contingency in a study, "Contingency and Natural Law," in 1970, as it was in Thomas Torrance's book *Divine and Contingent Order*.[14] The irreversibility of time with its consequence that every event is uniquely new and the indeterminacy of individual events according to quantum theory seemed to support the view that natural processes are basically sequences of contingent events, an assumption, however, that is not opposed to the operation of laws in their course because, in the unique sequence of contingent events, regularities may occur that can be expressed in formulas of natural law. The element of necessity in the concept of natural law is not in opposition, then, to the basic contingency in all natural events, though there is also unpredictable contingency and novelty. My argument received a critical response from Robert J. Russell.[15] At that time, Russell doubted whether contingency in the occurrence of actual events ("local" or "global" contingency) can reasonably be affirmed. But a quantum physicist like Hans-Peter Dürr, longtime assistant of Werner Heisenberg, looked at the world of quantum physics as early as 1986 in terms of an open system and wrote, "In a certain sense this world originates anew at every moment."[16] And the thermodynamic investigations of fluctuations in dissipative systems, which were pioneered by Ilya Prigogine[17] and others, showed that unpredictability and contingency occurs even in macrophysical processes. The observation of chaotic processes gave rise to a discussion on its deterministic and indeterministic aspects and even to a "chaos theology."[18] I agree with the judgment of Sjoerd Bonting that, notwithstanding the deterministic character of theoretical description, "The natural system becomes indeterministic at a bifurcation point."[19] I also agree with his theological defense of talking about divine "intervention" in such a situation.[20] I appreciate Arthur Peacocke's hesitation concerning such language in view of the danger of falling prey to another form of "God of the gaps." But if contingency is not an exception in natural processes, rather a basic character of all events, notwithstanding the regularities occurring in their course and described in terms of natural law, then it is no longer a matter of gaps, but of a different view of nature at large. The interaction of God with his creation is certainly concerned with creation as a whole because the eternal God looks upon creation from the point of its completion; therefore, there is an influence of the whole on the parts, "top-down causality." But God also relates creatively to every single creature, as Bonting emphasizes: "If we deny God's 'intervention' in his creation, we are back to the deistic God of Newtonian thinking."[21]

# ✢ 5 ✢

# Providence, God, and Eschatology

MODERN PROTESTANT THEOLOGY has been concerned with the theological interpretation of history for more than two centuries. Today it owes gratitude to Langdon Gilkey for his reminder that such a discussion should take due notice of the doctrine on providence, of the problems inherent in its classical presentations, and of its need for contemporary reinterpretation. As Gilkey has successfully demonstrated, the doctrine of providence offers opportunities for taking a fresh look at the problems of history and of its theological interpretation. One could be tempted, indeed, to cover by the notion of providence everything that modern theology has discussed in terms of a theology of history: the biblical history of salvation and revelation, including the Incarnation of Christ, the history of the church, and the eschatological completion of all human history. In principle, it could be argued that the notion of the divine government of created reality, in spite of the presence of sin and evil in the midst of it, comprises all those issues.

Langdon Gilkey decided to work with a more restricted concept of providence in line with the traditional notion of providence as God's preservation and government of creatures. Therefore, in his judgment, providence "must be supplemented by incarnation and atonement, and ultimately by eschatology."[1] On the other hand, it is precisely such a restricted notion of providence that serves as an instrument of correction in relation to the one-sided emphases of neoorthodox theology on christology and "contemporary eschatological theology" on the eschatological future. Gilkey does not want to deny the importance of any of these other emphases, but he wants to supplement them by the issues of providence and thereby to correct the dangerous consequences of their exclusivism. In the case of neoorthodoxy, Gilkey identifies its shortcoming in the fact that it leaves the world of social and historical reality to itself.[2] In the case of eschatological theology, he shows that the liberal concern for history was restored but that the developmental view of history as progressively approaching the kingdom of God was replaced by a prospectus of "radical social change in the name of a historical kingdom to come," with the acute danger of overlooking the ambiguities inherent in radical social change. In turn, radical social change can produce new forms of injustice, since sin is not a prerogative of clinging to the past but lurks in the corruption of human freedom itself.[3] In both cases—of neoorthodoxy and of the contemporary theology of hope—the category of providence is seen to be vanishing from the theological agenda.[4] Gilkey, on the contrary, considers providence as providing the necessary presuppositions for the doctrines of christology and atonement as well as of eschatology.

One may perhaps wonder whether the negative attitude toward social and historical reality, which Gilkey attributes to neoorthodoxy, was really characteristic of all its forms. It certainly was characteristic of Rudolf Bultmann's thought; and his individualism may have been the main reason for the sudden decline of Bultmannianism in the early days of the students' movement in the late 1960s. But the case of Barthianism has been different, since by analogical reasoning from christology Karl Barth reached quite definite conclusions concerning the political reconstruction of society. His sympathies for socialism, Marxist or otherwise, help to explain the resurgence of Barthianism in Germany, precisely during the years of the students' revolution. For the same reason, there is a line of continuity from Barth through the earlier Jürgen Moltmann to Latin American liberation theologies, while there was no such connection on the side of other theologians whose concern was focused on eschatology.

# Providence, God, and Eschatology

Since in characterizing eschatological theology Gilkey deals at some length with my own arguments in favor of a futurist perspective in theology, and especially in the doctrine of God, I may be forgiven [a] comment on his remarks.[5] While the reality of God, in my argument, is indeed bound up with God's kingdom so that God's activity in creation is seen as the ultimate future impinging upon everything present, this does not entail a "radical negation of the present order of the world."[6] I rather emphasized that, as the power of an ultimate future, God has been the future of all past events as much as of those still ahead of us, and I explicitly added that "we can now understand even our past as the creation of the coming God,"[7] since God's faithfulness grants continuity to creation. It is true that some theologians used the idea of the futurity of God in such as way as to oppose the future to the past and present, while overlooking the ambiguities of a future to be produced by human freedom, especially by political action. But that is by no means an inevitable consequence of the idea of God's futurity. It is rather a dualistic attitude, reminding one of the Gnostics of the ancient world. As long as the idea of one God as creative origin of the world is not surrendered, the futurity of God has to be understood as origin of the past and the present no less than of things to come, and the unity of God inevitably imposes a measure of continuity upon the transition from any past or present to the future. Therefore, the traditional issues of the doctrine of providence cannot be excluded from the concerns of such a theology.

When Gilkey argues against unlimited secularism regarding the interpretation of present and past human reality, every Christian who adheres to the belief in the one God, creator of heaven and earth, should applaud him. There is indeed "a dimension of ultimacy on every level of our being."[8] To substantiate this claim was the rationale of my endeavors in the field of anthropology as early as 1962 and again in 1983.[9] I fully agree with Gilkey's warning that "if God be solely the future, and not also the God of our present, this 'religious' dimension of social and political experience fundamental to that experience—as well as the mythical and demonic elements of politics itself—become unintelligible and irrelevant. . . ."[10] Therefore, I also share his criticism of Barth and Bultmann concerning the surprising fact that they accepted the "ontology" of the secular historical consciousness rather than trying to transform it.[11] It is not enough to relate a Christian interpretation of human life, personal and social, to contemporary experience in a merely extrinsic fashion, be it even by way of protest. Rather, theology has to enter into a dispute with purely secular interpretations of society and of personal life. The same applies to the interpretation of nature. A detailed argument concerning the "religious" or "ultimate" dimension in nature as well as in human life is indispensable, and it has to move on the same level as other philosophical reflections on the work of the relevant disciplines. Only on the basis of such an argument can theology hope for any degree of plausibility when it comes to statements concerning God's action in history.

While this level of general agreement concerns specifically the Tillichian heritage in Gilkey's thought, I find myself also in broad agreement with what he says in the line of Reinhold Niebuhr's analysis of the ambiguities of human freedom and its exercise in history. If I am not mistaken, Gilkey moves at this point beyond Tillich's language of estrangement toward Niebuhr's more Augustinian account of the human situation. Niebuhr's Christian realism seems particularly important in Gilkey's pointed criticism of the "apparent identification of liberation and especially of social and political liberation with the salvation promised in the gospel." Gilkey hits the decisive point when he comments on this position: "It is the corruption of freedom in ourselves, not the enslavement of our freedom to others that represents the most basic issue of history." Therefore, "greater self-determination does not guarantee greater freedom from sin. After all, present oppressors are precisely those who have in the past been 'liberated.'"[12] If there are some futurist theologies where such criticism applies, they certainly do not represent my way of arguing from eschatology.[13] Our bondage to the past consists precisely in the vigorous activity of the "old Adam" in our lives and in what we call our freedom, rather than following the spirit of the "new Adam" in ourselves. For this reason, I could not avoid disappointing some of my theological friends by refraining from revolutionary conclusions from the primacy of the kingdom. After all, Jesus did not join the zelotic dream of establishing the kingdom of God on earth by way of political liberation. Because of human sinfulness, no human society is possible without civil government; for the same reason, the noble task of civil government, the establishment of peace and justice, will not be perfectly realized by human administrators but only in the event of the transcendent kingdom of God. This need not rule out the possibility that human political action as well as political institutions may be inspired by the vision of the eschatological kingdom. But as soon as the condition of human sinfulness in this world is forgotten or, worse, attributed exclusively to the political opponent, the political activist no longer honors God and the kingdom as distinct from human efforts; and the vision of the kingdom gets distorted and perverted into the kingdom of one who puts himself in place of God.

Thus, I agree with Langdon Gilkey that eschatology must transcend the conditions of the present world and the possibilities of human action. Precisely in this transcendent quality the eschatological hope serves as a criterion for evaluating the present situation, but it also provides a source for illuminating and directing our ways through the history of this world. According to a famous dictum of Ernst Troeltsch, the transcendent empowers us to live our lives in the relativities of this world of finite existence.

But what end does it serve, then, to reformulate the doctrine of God on the basis of eschatology? Why talk of God in terms of "the power of future"? The main reason for doing so is—at least, in my mind—the need for a new ontological basis for Christian theology. This is another concern that I share with Langdon Gilkey.[14] Theology needs an ontological basis not only for its assertions on human nature and history but also and especially for the task of reformulating the Christian doctrine on God in view of the arguments of modern critics of traditional philosophical theology and, in this context, also of modern atheism. With Gilkey's fine account of Augustine's doctrine of providence and its underlying concept of God (as opposed to basic concerns of the modern mind that are expressed in modern science as well as in modern history and philosophy), my judgment converges that the classical Christian concept of God, if rigorously developed, could result in unacceptable, deterministic consequences. This insight occurred to me when I studied the medieval doctrine on predestination and divine prescience in order to prepare my doctoral thesis in the early 1950s. There are, of course, many other problems with the traditional doctrine of God's essence and attributes, some of which are connected with the more general problems of the traditional metaphysics of substance. But neither the reformulations of the concept of God by the philosophies of German idealism, especially by G. W. F. Hegel, nor the attempts of process philosophers, especially Alfred North Whitehead's doctrine of God, can serve as an acceptable substitute for the traditional concept of God in Christian theology. I share Gilkey's reservations with regard to Whitehead's and Samuel Alexander's concepts of God, because Christian theology has to conceive the reality of God as "the source of our total existence."[15] On the other hand, the process philosophers' criticism of the traditional ontology of substance has to be taken seriously, especially in its convergence with the call of other thinkers like Martin Heidegger or Ernst Bloch for revising the separation of being from time.[16] But as there is no generally accepted or theologically satisfactory solution of the ontological problem available, theologians themselves have to enter into the philosophical dispute at this point, although they may not be in a position to work out a comprehensive and detailed treatise on metaphysics. Such was my situation when I wrote the first chapter of *Theology and the Kingdom of God*, trying to reformulate the concept of God from the point of view of Jesus' eschatological message and his emphasis on the priority of the kingdom over any other concern.

This attempt was located in the context of the discussions on being and time and was related to their impact on the foundations of the traditional doctrine of God. It was not possible, of course, to discuss in that connection the whole range of the related problems of ontology, although a hint in that direction was given by the essay on appearance that was added as an appendix to *Theology and the Kingdom of God*. Meanwhile, in connection with the development of my own systematic account of the doctrine of God, I obtained more clarity on a number of related issues in the context of the history of metaphysics; and an invitation from the Institute for Philosophical Studies at Naples provided an opportunity to write down a series of lectures on metaphysical questions involved in the reformulation of the concept of God.

# Providence, God, and Eschatology

While Gilkey and I do not disagree about the basic requirement of ontological reconstruction for the task of theology, the direction that we envisage for that reconstruction may be somewhat different. In his assessment of the contrast of modern historical consciousness to the presuppositions of traditional Christian doctrines on providence, Gilkey starts with the new sense of historical and cultural relativity that characterizes the modern mind. He goes on to describe the new views of human creativity and freedom and then considers the "temporalization of being" as "indebted more to the developing themes of historicity and autonomy than to that of law in the modern historical consciousness."[17] Of course, I agree that the sense of historical and cultural relativity is distinctively modern, but I do not think that the emphasis on autonomy arose as a consequence of this, nor do I see the temporalization of being arising as a corollary to the modern conception of freedom. To consider time primarily in terms of actuality and possibility of decision, as Gilkey does,[18] occurs to me as a somewhat unsatisfactory account of the nature of time, although I grant that, in relation to the human awareness of freedom, the difference between present and future may appear in such a fashion. Gilkey himself says that temporal passage is to be conceived as "the ground of creativity." But how, then, can it become this as a consequence of entities and events being self-creative?[19] Does it not have to be the other way around, if temporal passage is to be conceived as the *ground* of finite creativity? But then the nature of time itself has to be described in different terms, unless we end up with a circular definition.

In fact, the temporalization of being developed from different roots than the autonomy of decision. The Kantian doctrine described time as a subjective form of experience in relation to self-awareness rather than to autonomy. Later on, Heidegger could even refer to Kant's doctrine on time as the basis for his criticism of modern subjectivism. Henri Bergson's analysis of time focused on the experience of "duration," not autonomous decision. He used his philosophy of time as a starting point for a redefinition of freedom, not the other way around. In Wilhelm Dilthey's analysis of historical experience, time was conceived not in terms of decisions but of occurrences, a sequence of contingent events which continuously change the frame of reference in our awareness of meaning and significance. It was only in Whitehead's metaphysics that the ideas of subjectivity and decision became ontologically basic in the description of time and that, therefore, the future became a mere possibility. Gilkey's description of time as "the movement of events from possibility to actuality" makes sense only on the basis of Whitehead's concept of creativity. But the description is circular, since it identifies time only in terms of the word *movement*, which already presupposes the idea of time. The insight that time cannot be reduced to movement resulted already from the discussion on the nature of time in classical antiquity.

Moreover, Gilkey criticized Whitehead's doctrine of creativity on theological grounds, and at least to me his objections seem valid and theologically decisive. But if creativity is to be reconceived in terms of God's activity as creator and preserver (see note 15 above), then it does not seem consistent to use Whitehead's idea of self-determination of events as a basis for describing the nature of time. Rather, it will be inevitable to conceive the mode of God's being as constitutive of everything created, including time. That is to say, Christian theology cannot and should not try to avoid the notion of eternity as constituting the nature of time. This does not preclude a redefinition of eternity itself, especially a revision of the concept of eternity as fundamentally opposed to time. It may be that Augustine's idea of eternity was deficient in this respect. But if so, it was deficient in his own time already, since Plotinus had developed a theory of time and eternity that conceived of eternity in terms of the totality of life that, in the perspective of time, appears as a sequence of separate moments. The unity of time itself, according to Plotinus, depends on the idea of eternity. Only the future can recover the wholeness of life in the sequence of temporal moments. Unfortunately, Augustine did not recognize the potential of this coincidence of eternity and the future of time for a Christian eschatology and for his own theology of history. Nevertheless, the fault does not lie with his idea that everything is present to God's eternity.[20] Time and space are bound up with finite entities, and from every finite point of view, they are different. But God, in his eternity, does not exist under the limitations of time and space. When I speak of God as the power of the future, I mean that the future of finite entities is the point where time and eternity coincide, the "place" also for that eternal presence that ancient Greek thought conceived as timeless, but that, in fact, in the course of time and history is available only by anticipation. This is not a simple spatialization of time, because the future of every finite entity occurs as contingent. The difference between a contingent future and the present seems to provide the main reason for the irreducible difference between time and space. But Bergson has been justly criticized for his almost complete separation and opposition of time and space. No satisfactory philosophy of nature is possible that does not also account for the interconnection of time and space.

If there was no fault in the traditional assertion that everything finite is present to God in eternity, the case is different with purposive action that can be attributed to God only metaphorically because, in the instance of eternity, there is no separation between goal and execution. This applies also to predestination. The metaphor of purposive action pictures God at the beginning of the world process as looking ahead and determining everything from the outset. This anthropomorphic image of God's relation to the world, taken literally, was mainly responsible for the clash of the traditional doctrine of providence and predestination with the experience and intuition of human freedom, the point where Gilkey rightfully identifies one of the basic differences between modern historical consciousness and the traditional doctrine on providence. Those difficulties do not occur when God is understood to act from the eschatological future, because the notion of finite freedom is opposed only to the idea of determining forces that arise from the past. Admittedly, the ontology underlying and substantiating the argument of eschatological theology has not been worked out in detail, and the amazement that Langdon Gilkey shares with others in relation to talk about God as acting from the future is understandable.[21] But such language expresses a way of dealing with the ontological question of how the action of God in relation to contingency in general and to human freedom in particular can possibly be understood. One should expect similar ontological proposals from Gilkey himself, if his rejection of Whitehead's assumption of a principle of creativity besides God would be followed through consistently. All creativity then is to be attributed to or derived from the creator God, and at the same time the deterministic tendency of the traditional doctrine of providence is to be avoided.

I agree with Gilkey that the very act of creation involves some "self-limitation" on the part of the creator God in the sense that creation means that an existence of their own is granted to the creatures. Since this is a logical requirement inherent in the idea of creation, it applies to all conceivable ontological models of God's relation to a world of God's creation. For the same reason, to be sure, I am not happy with the word *self-limitation* because the act of creation, with everything it implies, is to be considered as an expression of God's sovereign will. But that may be a purely terminological problem. It seems more important that we agree concerning the ontological requirement of some autonomous existence of the creature as entailed in the idea of creation. But, of course, this does not relieve the theologian from the task of accounting for the kind of causality that is bound up with any idea of a sovereign creator God, nor does it actually limit the responsibility of the creator God for the world God created.

I understand Gilkey's argument that the question of theodicy is not yet solved simply by moving the conception of God's determinative influence upon creation from its beginning in the past to its future. Such an operation indeed does not solve the problem of theodicy. But it seems to me a precondition for any acceptable answer to that question. The idea of divine self-limitation as entailed in granting the creatures some existence of their own does not itself answer the question whether it was worthwhile or even responsible to create a world that would involve so much perversion and suffering. It does not even provide a basis for answering that question. Only the Christian belief in the reconciliation of the world in Jesus Christ and the hope for its final eschatological consummation can do justice to this complaint because it points to a future when God will "wipe away every tear from their eyes" and put an end to mourning, crying, and pain (Rev. 21:4).

The ontological basis that Langdon Gilkey postulates for a reformulation of the doctrine of providence is particularly important in reinterpreting the notion of preservation. Gilkey does not go into much detail here, but it seems obvious that this issue requires a great deal of work because, since the introduction of the principle of inertia in seventeenth-century physics, the old doctrine that each creature needs a continuous action of God to preserve its existence has lost its plausibility. Such an external preservation became superfluous. To my knowledge, the Christian doctrine of providence never recovered from this blow, and very rarely has the issue been faced. Here the dialogue with science, especially physics, seems indispensable for any hope to overcome this impasse.

There is one other issue that I miss in Gilkey's treatment of providence. That is the notion of election. It can be argued, of course, that the doctrine of providence is not the proper place to deal with election. In my own systematic theology, election will be treated in connection with ecclesiology because a Christian doctrine of election seems to require a concept of the church. But in the history of the doctrine of providence, the issues of election and predestination were closely connected with providence. Even if, in the present situation, the doctrine of election and predestination have to be revised in terms of a continuous action of God in history, one might contemplate whether some more general treatment of this subject would be appropriate in connection with God's governance of the course of history.[22] Gilkey deals with these matters in terms of the tension between destiny and freedom and especially in relation to how destiny becomes fate. He made many pertinent observations on this matter. But his development of the doctrine of providence into a theology of history could profit from including also some general consideration on election. Is not the notion of election related to the issues of creative possibility that is provided by God to creatures and that has been deservedly emphasized by Gilkey?[23] If possibility is conceived not only in terms of formal alternatives but in terms of a lure or of some special calling that is accompanied by a corresponding element of responsibility, then the notions of vocation and election seem rather close at hand. Their inclusion could also serve to broaden Gilkey's concept of "destiny," which he relates primarily to the "given" in personal and social life that is inherited from the past.[24] But if destiny is also related to our identity, as Gilkey affirms, does it not then include a relation to future possibilities, to a future completion of our identity? Is our identity not constituted by what we feel ourselves "meant" to be? If so, some notion of election should be included in the awareness of destiny. This would also be coherent with the fact that the transformation of destiny into fate, which Gilkey so eloquently addresses in many places, corresponds to the traditional notion of judgment.

Langdon Gilkey's work in theology gave me the rare pleasure of reassurance that there is a broad basis that theologians of different opinions can share in the assessment of the rich tradition that modern theology inherited from the past together with a clarification of the intricate problems inherent in the traditional doctrines. On this basis, there can be a high degree of convergence in the evaluation of the challenges that modernity poses both to traditional doctrines and to the contemporary work of theology. It is this sort of framework that allows for serious discussion of different opinions and for the development of a culture of theological discourse.

# ⇢6⇠

# A Dialogue

*God as Spirit—and Natural Science*

IN THE DIALOGUE between theologians and scientists, it is important to be aware of the fact that such dialogue does not move on the
level of scientific or religious discourse but rather on the level of philosophical reflection on both scientific terms and theories and religious
doctrines. Usually, when scientists talk about the general meaning and
significance of their equations and theories, they already move on some
level of philosophical reflection. I say "some level" because such talk
does not always exhibit the same degree of philosophically instructed
reflection. The philosophical sophistication may be rather poor; still the
scientist, the authority of his or her scientific competence notwithstanding, argues on the level of philosophical reflection when addressing the
public on the broader significance of his or her work in science.

In the dialogue between science and theology, the fact that both
sides meet on a level of philosophical reflection is of particular importance. Such a discussion requires a rather high level of philosophical
sophistication because the traditional doctrine of God is related to
many philosophical issues that imply some relation to the language of
science. This is the case in the notions of causality, law, and contingency. These notions are indispensable in any discussion about God's
action in the world. Similarly, the presence of the transcendent God in

the world he created has to be considered in connection with the concepts of space and time. Otherwise, talk of God's omnipresence would be empty. Similarly, the clarification of the concept of divine eternity requires a discussion on how it is related to temporal events.

The contingency of events and the relationship of the concept of contingency to that of natural law was the focus of a debate in a German group of physicists, philosophers, and theologians in the 1960s. My article on "Contingency and Natural Law," written for these discussions, was first published in 1970 and appeared in English translation in 1993 in a book edited by Ted Peters (*Toward a Theology of Nature: Essays on Science and Faith*). During those conversations in Germany, the issue of contingency was considered to be fundamental in both disciplines, science and theology, though in different ways. In science, the concept of laws of nature presupposes initial conditions and boundary conditions for the application of a particular law in the description of natural processes, and these preconditions are contingent relative to the formula of law. Although these initial and boundary conditions may be explained by another law, this again presupposes contingent conditions for its application. This fact could be understood on the assumption that basically all events occur contingently, but with some degree of uniformity in their sequence, a uniformity that lends itself to a description by hypotheses of natural law.

In Christian theology, the contingency of events characterizes the logical form of God's actions in history. Contrary to its Aristotelian origin, the theological concept of contingency emerging in the Middle Ages corresponds to the freedom of God in his creative action in history. This concept of contingency requires, of course, more than the nomological contingency of initial and boundary conditions relative to a given formula of natural law. It indicates the contingent occurrence of actual events, be it in particular cases (local contingency) or in all events (global contingency). At the time of the first publication of my article, contingency of actual events seemed to be suggested primarily by the unpredictability of individual events in quantum physics but could be doubted with regard to macrophysical processes as well as with regard to the regularities of quantum mechanics. In 1988, Robert Russell expressed such doubts in a critical article published in *Zygon*.[1]

Since that time, however, the development of chaos theory seems to have shown that "local" contingency of actual events does indeed occur even on the level of macrophysical processes, without precluding the description of such processes by natural laws. This supports, in my judgment, the assumption that basically all natural events happen contingently, notwithstanding the uniformities that ordinarily occur in their sequence, and permit a description by formulas of natural law.

In my judgment, the basic contingency of each and every event in the world of creation, including the occurrence of order and uniform pattern in the sequence of events, is much more fundamental concerning the task of a theological interpretation of the natural world in terms of creation than the idea of purpose is. In connection with the discussion of an "anthropic principle" guiding the development of the universe from the beginning to the emergence of life and of intelligent life, the old idea of a teleological determination of physical processes (a determination aiming at the end of the process) has acquired new plausibility in the eyes of many. The belief in a teleological orientation of the development of the universe, as founded in its beginnings, seems to lend itself more easily to the assumption of a divine purpose governing that development. But for two reasons I hesitate to follow these suggestions, one of these reasons being scientific and one theological. The assumption of an intrinsically teleological direction of the development of the universe requires acceptance of not only the "weak" anthropic principle, according to which the later emergence of life depends in fact upon the fine tuning of the natural constants in the early phases of the universe, but also of the "strong" anthropic principle, which claims that the early condition of the universe renders the later emergence of life and of intelligence inevitable. This strong anthropic principle does not seem exactly empirically warranted, and it is in conflict with the role of contingencies in the history of the universe.[2] On the theological side, it is certainly possible to speak of a "purpose" of God in creating the universe with reference to the creation of human beings and to their final redemption in the eschatological future. It is possible, in theology, to talk that way because God's act of creation relates to the universe as a whole, including its later developments. Thus, the beginnings and intermediate stages can be contemplated in the light of the outcome toward which they lead. But still, the language of "purpose" easily suggests a false anthropomorphism in our language about God because it suggests a position of the Creator in the beginning of the universe as if looking ahead to a distant future and selecting means for achieving some purpose. Such a picture ignores the eternal presence of God, to whom the future is not distant and who, indeed, precisely in his capacity as the final future of everything, is the source of ever-new contingent events. The language of a divine "purpose" realized in the development of the universe is legitimate only with regard to the fact that the divine act of creation relates to the universe as a whole and therefore includes its final future as well as its beginning. Indeed, the character of the universe as a whole is "determined" by that final future, and the power of the future completing the history of the universe can also be understood as the source of contingent events that spring from that future during the entire course of the universe's history.

But how is the Creator to be understood as active in the particularities of his creation? This question is not yet answered by the assertion of the contingency of all events and of all more or less enduring forms of created existence. How can the Creator be understood with the apostle Paul as *dynamis,* force that is at work in his creation (Rom. 1:20)? And how is God's divine power related to the natural "forces" that are effective in the movements of his creatures? As these movements are taking place in space and time, God's relationship to space and time has to be clarified in such a way that his powerful presence with his creatures and their movements in space and time becomes intelligible.

In Isaac Newton's doctrine of space, absolute space was conceived as a medium of God's presence at the place of the finite existence of his creatures. God was understood as mind, who by his will is present and active in the material universe as our soul is present in all the parts of our body. The biblical basis of this conception was John 4:24: "God is spirit." For many centuries, since the work of Origen in the third century, this sentence had been interpreted in the sense that God is mind, *Nus,* and Newton presupposed that tradition. But in the Bible spirit does not mean "mind." The Greek word *pneuma* as well as the corresponding Hebrew word *ruah* rather mean "wind," "storm," or "breath." Thus, in the biblical creation story, when it is said that in the beginning

God's "spirit" was "moving" over the primeval waters, the image is that of a storm agitating those waters. This spirit is the source of all movement. It is also the source of life in animals and in human beings. In Genesis 2:7, it is said that God breathes his spirit into the nostrils of the figure of Adam that he had formed from clay, and in Ecclesiastes 12:7, we learn that in the moment of death "the spirit returns to God who gave it." In Psalm 31:5, we read, "Into thy hands [God] I commit my spirit," and according to Luke 23:46, Jesus used these words on his cross immediately before he died: "Father, into thy hands I commit my spirit." Thus, the life we received from the breath of God is within us until our last breath. The spirit as divine wind or breath is very important in the biblical understanding of life and movement. But it seems a world apart from modern science and from its ways of accounting for the movements of bodies and especially for life.

So I thought when I happened to read an article by a renowned historian of scientific terminology, Max Jammer, on the concept of *field*.[3] Jammer, who has published important books on the concepts of space, of mass, and of force, characterizes in his article the modern scientific concept of field as a further development of the ancient Stoic doctrine of *pneuma*. Jammer even called the Stoic *pneuma* the "direct precursor" of modern field theories. Now, the Stoic concept of *pneuma* was in many ways similar to the meaning of the biblical word *pneuma* or, in Hebrew, *ruah*. The basic intuition was in both cases that of air in movement, full of force, which, according to the Stoics, is a result of the "tension" the air contains. The important difference, of course, is that in Stoic philosophy the *pneuma* was a cosmic principle, pervading the cosmos and keeping all its parts together by its tension (*tónos*), while in the biblical conception the divine *pneuma* was conceived as transcending the world of creatures though working in it creatively. Otherwise, however, the biblical and the Stoic conceptions were similar. This similarity, together with the observation that the Stoic concept of *pneuma* was the "direct precursor," as Jammer put it, of the modern scientific field concept, suggested the conclusion that the meaning of the statement in John 4:24 that "God is spirit" is considerably closer to the field concept of modern physics than to the Platonic idea of a divine mind or *Nus*, which Origen identified with the biblical concept of God as *pneuma* because he

abhorred the materialistic interpretation of *pneuma* by the Stoic philosophers. Origen made fun of the bodily nature of the Stoic *pneuma* by saying that bodies can be divided and put together again, which contradicts
the basic requirements of any concept of God as first principle because
all division and composition is in need of a further cause doing the job
of the dividing and composing. This argument secured the acceptance of
Origen's identification of the divine Spirit with *Nus* for many centuries,
though that identification was not correct as a translation of the biblical word *pneuma*. But precisely at this point, modern field theories offer
the theologian a conceptual help because the spreading of field effects is
no longer considered to be dependent on a material medium like ether
but requires only space or, in the general theory of relativity, space-time.
Thus, the biblical conception of God as *pneuma* can be defended against
the suspicion of involving some bodily conception of God without iden-
tifying *spirit* with *Nus* or mind.

Against this, John Polkinghorne recently argued that "this notion
of a field's immateriality is not correct," since notions like energy and
momentum "function in the same way for the field as they do for particles of matter."[4] Now, the concept of matter is not—as "mass" is—a
strictly physical concept but a philosophical one, and there seem to be
different opinions among physicists as to the impact of modern physics on the formerly "materialistic" character of physics. The German
theoretical physicist Georg Süssmann, for example, argued that there
has been a change in that respect so that modern physics should no
longer be called materialistic.[5] Albert Einstein himself distinguished
the "field of gravitation" from "matter."[6] My own point was that the
fields of modern physics are not bodies in the way the Stoic doctrine of
*pneuma* considered *pneuma* a bodily reality, occasioning thereby Origen's objection to its application in the doctrine of God. Like the ideas
of dynamism in modern philosophies of nature since Gottfried Wilhelm Leibniz and Ruggero Boscovich, the introduction of field concepts into modern physics since Michael Faraday involved the idea of
a priority of the field with regard to bodily particles.[7] Thus Max Jammer wrote, "For a consistent field theory the concept of a 'particle' is
extraneous. It seemed therefore very tempting to interpret mass points
as singularities of the potentials of the field equations."[8]

Another objection to my application of the field concept in the interpretation of the biblical language about God as spirit has been that it does not use the word *field* with the precise meaning it has in physics.[9] That is certainly correct. I do not contend that the divine Spirit is sending forth waves that can be counted and measured. But neither is the word *field* as applied to God, who is spirit, just a vague analogy or a poetic expression. It is certainly a metaphor, as the field concept of physics itself is, because the primary meaning of *field* is the field of the farmer, where wheat or corn is raised. The origin of field language in the sciences, then, is certainly metaphorical.[10] But it is not a vague analogy, either in science or in theology. It has a clear conceptual meaning in its connection with the concepts of space and time. If that were not the case, the use of the field concept would indeed become vague. It is because of its connection with the concepts of space and time that a sufficiently precise theological use of the field concept is possible that is clearly distinct from its use in physics and yet related to it. The reason is that space is the minimal requirement for any notion of field, while it may dispense with the idea of force as in general relativity. Since in physics the notion of field is connected with that of energy, in addition to space the dimension of time is needed, or space-time in the case of general relativity. In theological use, talk of God the spirit in terms of field also implies a connection with the concepts of space and time, though different from their use in physics. This affirmation, of course, needs some explanatory remarks, and so I offer some considerations on space and time before returning to the field concept.

The question of how the eternal God is related to space and time has a long history. The idea of God's omnipresence always required some such connection. Although in his eternity God transcends time, still he is present and becomes present in the temporal reality of his creatures. In the early eighteenth century, the function of space in God's omnipresence with his creatures became the object of a famous controversy between Gottfried Leibniz and Samuel Clarke, who, on behalf of his friend Isaac Newton, defended Newton's occasional reference to space as *sensorium Dei* (sensory) against the suspicion of pantheistic implications. Leibniz had used the argument of Origen against the Stoics that, if space were an attribute of God, God had to be composed of parts and would be divisible into parts. Clarke's rejoinder was that geometrical space that consists of parts presupposes some undivided and infinite space because every act of composition or division already presupposes space within which the dividing or composing takes place. This infinite and undivided space is, according to Clarke, identical with the divine immensity. It is presupposed in our conceiving of parts of space and in any composition or division; hence, that undivided infinite space is also prior to measurement, which presupposes and applies standard units that are already parts of space. All geometrical conceptions of space, then, because they operate with units of measurement, already presuppose the infinite whole of undivided space. According to Clarke, that infinite and undivided space is the divine immensity, the field of God's omnipresence, in distinction from the geometrical space, where we have parts, composition, and division and which is also the space of the physicist and of his measurements. If one identifies the space of geometrical measurement with the divine immensity, as Benedictus de Spinoza did, one ends up with pantheism, but Clarke (and Newton himself, as Clarke believed) clearly observed the distinction between God and the infinite and undivided space of his immensity, on the one hand, and the geometrical space of the physicists' descriptions of the world of nature on the other.

At the end of the eighteenth century, Immanuel Kant, in his *Critique of Pure Reason* (1781), repeated the argument of Clarke that every conception of partial spaces or space units presupposes the intuition of one infinite and undivided space, within which we may conceive of circumscribed space units. The same with time: The perception of any section of time presupposes an awareness of time as an infinite whole. In the traditional philosophy of time, such simultaneous presence of the whole of time was called eternity. Plotinus said in *Enneads* III, 7 that the perception of the present moment and of temporal continuity in proceeding to the next moment is possible only on the basis of some awareness of eternity, which is to say, the simultaneous presence of the whole of life. Kant, in his later period, no longer paid attention to the theological implications of our awareness of space and time because he was concerned with avoiding pantheistic affinities, but he kept insisting that, in both cases, awareness of the infinite whole of space and time is presupposed in the perception of any part of time or space.[11]

If all measurement in space and time presupposes the undivided infinite space of God's immensity and the infinite whole of his eternity, then the definition of the concepts of space and time cannot be the exclusive prerogative of the physicist and the mathematician. Their special competence is the measurement of space and time, but in exercising that competence, scientists move within an intuitively present conception of space and time that is neither constituted nor exhausted by their measurements. The question of the nature of space and time, therefore, transcends physics and geometry. This is also the reason why changes in the scientific description of time and space like the space-time concept of the general theory of relativity contribute less than one might think to the philosophical question of the nature of space and time.[12] The contribution of physicists concerns the measurement of movements in space and time, which is important enough; but contrary to Spinoza and Einstein, the nature of space and time transcends any geometrical model of their description because the infinite and undivided whole of space and time—or of space-time, for that matter—precedes all measurement. God's immensity and eternity are prior to the finite reality of the world of creation that is the object of geometrical construction and of physical measurement. The infinite space of God's immensity, however, and the infinite whole of simultaneous presence that is God's eternity are implicated and presupposed in our human conceptions and in our measurements of space and time. Thus, God's eternity is different from the time of his creatures but constitutive of it, and his immensity is constitutive of the space of his creatures. This comes to expression in that the infinite whole of space and time precedes any conception of temporal and spatial sections or units and all geometrical description.

At this point, I return to the field concept and to the significance of its application to the doctrine of God as spirit. I said before that space and time, or rather space-time, are the only basic requirements of the field concept in the general theory of relativity. Here, the universe is described as a single field, while, in principle, the states of bodily matter (or particles) are considered as singularities of the cosmic field. If all geometrical descriptions of time and space, however, are dependent on the prior conception of space and time as an infinite and undivided whole, the immensity and eternity of God, then this infinite and undivided whole may also be described as infinite field, the field of God's spirit that constitutes and penetrates all finite fields that are investigated and described by physicists, even the space-time of general relativity. This relationship makes intelligible how the divine Spirit works in creation through the created reality of natural fields and forces. The interpretation of the concept of God as spirit in terms of the field concept, then, functions as a key to obtaining some understanding of God's fundamental relationship to the world of nature.

Such a theological use of the field concept does not and need not rely on any specific field theory that physicists have produced.[13] Nevertheless, it is related to the field language of physics because it claims to deal with the preconditions of any physical field that occurs in the spatial and temporal setting of the universe. John Polkinghorne, in his criticism of my theological use of field language, did not pay any attention to my argument concerning the connection of the field concept with the concepts of space and time. Otherwise, he would have seen that the field concept is used not in a vague way but with some "precise meaning," as he demands, though different from the field theories of the physicists.[14] When I referred to Faraday, it was more the metaphysical vision behind his scientific work, the priority of his field of force regarding bodily entities,[15] that I was concerned with, though I am aware that the concept of force tended to be eliminated later on, especially in the field theories of Einstein.[16] When I called upon Einstein's field concept, I was primarily interested in its reduction to space-time, but I did not adopt the idea that the nature of space and time is determined by measurement and expressed in the geometrical scheme of space-time. I rather suspect this position is related to Einstein's professed Spinozism, and I think that theology has to insist on the transcendent reality of God even with regard to his immensity and omnipresence. My theological considerations on the divine Spirit as field aim precisely at the distinction as well as interconnection between the reality of God and the world of nature concerning its constitution in time and space. The space and time of creatures are composed of parts and are divisible into parts, which God is not, and they are objects of geometrical description and physical measurement, which God isn't either. It seems that the transition from God's immensity and eternity to the space and time of creatures occurs when finite events and entities are granted an existence of their own within the undivided space of God's omnipresence and in the presence of his eternity. The finite existence of creatures entails relationships that are described in schemes of measurable space and time. The space-time concept of general relativity is of philosophical significance here in expressing the dependence of the metrical structure of space and time on the presence of finite reality, of "masses" or clusters of energy. It is possible to conceive of this dependence in reverse by reducing the occurrence of masses completely to the concept of space-time.[17] But the relevance of these attempts seems to be limited by the irreducibility of events in quantum physics. This is related to the emphasis on contingency discussed earlier in this presentation.

In his critique of my theological use of the field concept, John Polkinghorne denied "that fields as such have any intrinsic relationship to contingency."[18] That depends on how the concept of field is formed. Polkinghorne's judgment is certainly correct with regard to classical field theories in physics, but he himself allows for an exception in the case of quantum fields. The field concept in general should make room for contingency, however, if time is to be taken seriously as a source of novelty that characterizes each new event because of the irreversibility of time. Whether such an accommodation of the field concept in order to describe the openness of natural processes is possible in physics along the lines, for example, as suggested by Ilya Prigogine, is not for the theologian to decide.[19] But it is certainly appropriate with regard to the work of the divine Spirit who is the creative origin of life in all its forms. Life is characterized by self-transcending openness in the case of the individual organism as well as in the case of the evolution of living forms. The ecstatic openness of life to its environment and to its future corresponds to the creative activity of the divine Spirit; and if the divine Spirit works as a dynamic field, then here we have a field concept that is connected with contingency regarding the efficacy of the field.[20] It also produces the phenomena described by chaos theory and is related to "the spontaneous generation of large-scale orderly structures in complex systems" as well as to the "effects of wholes over parts"[21] that Polkinghorne and Arthur Peacocke call top-down causality. When we want to describe the emergence of these phenomena in the language of a Christian theology of creation, we have to speak of the activity of the creative spirit of God in cooperation with the divine *Logos*.

In the Bible, references to the Divine occur in different ways. One way is the identification of the essence of God as *pneuma* as it occurs in John 4:24. More often, the *pneuma* is seen as the power through which God is active. Furthermore, the *pneuma* is mentioned as a gift of God in the hearts of believers or finally as a hypostatic reality of its own, glorifying the Son and the Father. When it is said that "God is *pneuma*," however, it must be added that God is not only *pneuma*. He is also a personal reality, more precisely a threefold personal reality. The divine Spirit exists in personal centers, in the Father, the Son, and the hypostasis of the Holy Spirit. Perhaps we may say that the field of the divine Spirit has three singularities—Father, Son, and Holy Spirit—and that it exists only in these three singularities, though radiating through all the world of nature, their creation. In all of his creation, God the Father is working through his Word and through his lifegiving Spirit.[22] Both can be related to the scientific description of the world of nature, the divine *Logos* as creative origin of the forms of creatures and of their order, the Spirit with regard to the dynamics of the natural processes. The recognition of the nature of spirit as field, in connection with a theological appraisal of space and time, contributes to elucidating this affirmation.

PART THREE

# RELIGION AND
# ANTHROPOLOGY

# ⇥ 7 ⇤

# Religion and Human Nature

Aᴍᴏɴɢ ᴛʜᴇ ᴜɴɪǫᴜᴇ characteristics that distinguish the human being from the most closely related of the higher animals is the fact that human beings worship gods or divine powers in some form: that is, they have religion. This circumstance is as characteristic of what is specifically human as are the ability to speak and the use of fire and tools. Nevertheless, the humane sciences and philosophical anthropology treat this theme only marginally, if at all.

We can distinguish three attitudes toward the subject of religion. The first suppresses it altogether: thus, Claude Lévi-Strauss, in whose treatment of totemism the religious character of the phenomenon vanishes and is reduced to a primitive and awkward preliminary form of abstract thought. According to Lévi-Strauss, totemism is only a first attempt at constructing general concepts, nothing more. He explicitly criticized Marcel Mauss for considering the possible reality of a numinous power, a *mana,* in the experiences of primitive peoples. Such notions, he believed, must be reduced to their functions.

The second attitude treats religious phenomena not as illusions that must be resolved by anthropology but still as marginal phenomena expressing a basic structure of human reality that can be described in different, purely secular language. One example of this is Helmuth Plessner, who described human uniqueness in terms of the concept of an eccentric life-form. This means that human beings can take a position outside themselves and evaluate themselves from without. Thus, human beings have self-awareness. But because human beings can distance their eccentric life-form from each individual experience and attitude, they are oriented towards "an open surplus": "Only an All, an extreme of power and majesty, can counterbalance this openness and give it an adequate support."[1] Religion is not called an illusion here, but neither is it regarded as constitutive for the structure of human life. It corresponds to a structure of humanity that in and of itself need not be described in religious terms in order to be understood adequately. Similarly, Arnold Gehlen, in his interpretation of totemism, regarded religion as a basis for the existence of institutions independent of individuals. For Gehlen, institutions constitute the counterbalance to the human openness to the world, something necessary to the stabilization of their attitudes; what gives human beings the requisite support, at least at the beginnings of humanity, is religion. However, unlike Plessner, Gehlen sees religion only as an expression and example of the nature of institutions as ends in themselves, which can also be described in purely rational terms. Therefore, Gehlen seems not to have regarded religion as an enduring condition for the functioning of institutions.

The third strategy for dealing with the fact of religious life, especially in the early history of humanity, treats religion quite explicitly as a transitional phase in human development. So according to Émile Durkheim and Jürgen Habermas, who took up Durkheim's thought and developed it further in his theory of communicative action, religion is, in fact, fundamental to the life of primitive humans, especially for their social systems. But the power of religion is only representative of the power of society over the individual. Hence, the enlightened human being no longer needs religion. For him or her, theology is replaced by sociology. This opinion can, of course, give way to the first, the notion that religious consciousness is pure illusion. Lévi-Strauss was a product of Durkheim's school.

All three of these positions assume that human beings have always been just as secularly constituted as today's social scientists see them to be. But they also produce religious ideas, either as a correlate of their self-transcendence, which reaches toward the indeterminate, or as a primitive form of secular concept-building, or as the expression of the power of society over the individual.

However, the assumption that from the beginning, by nature, human beings can be understood as purely secular beings contradicts the material that has been uncovered by paleontology, ethnology, and cultural history. An overwhelming wealth of findings attests to the constitutive significance of religion, especially at the beginnings of human development and in early cultures. If, in interpreting these findings, one assumes that the forms of human life can in principle be described in secular terms, one is, so to speak, rubbing against the grain of the materials of the cultural tradition. One can certainly do that, but one should not underplay the burden of proof associated with such a procedure, which lies entirely with the modern interpreters. It could, after all, be the case that it is not the whole premodern cultural history of humanity but the secular mentality of modernity itself that has fallen prey to illusion on this point, an illusion that may be bound up with the origins of the modern humane sciences in emancipation from the religiously shaped culture of Christianity.

Let us now turn to the indicators that point to an original association between humanity and religion. I will limit myself here to three:

1. First, there is the antiquity of burials. According to Karl J. Narr, the appearance of burials of the dead is always to be seen as an indicator of faith in a "life beyond death in some form or other."[2] Similarly, Anthony Francis C. Wallace finds that, wherever burials are found to have occurred, religion proves likewise to be present in some form.[3] The burials that have been found go back to the early Stone Age, and Narr supposes that the custom of burials must be still more ancient because, while burials in that early period were only in caves, early people normally lived in the open fields and constructed dwellings. In any case, in 1974 Narr used the demonstrability of burials as a decisive criterion for determining the end of the transitional phase from beast to human and thus for the beginning of humanity. That means that religion and humanity belonged together from the beginning, in fact as the decisive criterion for separating the human from the prehuman.

2. The second finding to be mentioned here is the religious basis of all cultures. This fact is generally recognized by anthropologists,

despite criticisms such as those Bronislaw Malinowski directed at Émile Durkheim and Lucien Lévy-Bruhl, because Malinowski only wanted to affirm the relative independence of secular spheres of life and their ordinary practices apart from myth and cult, even in early nonwriting cultures, in contrast to Durkheim and Lévy-Bruhl's pansacralism. On the other hand, Malinowski himself emphasized the fundamental significance of myth—and thus also of religion and its cultic practice—for the unity of ancient cultures. So also, according to Elman R. Service, the origin of political governing structures cannot be explained by purely economic needs—let alone the thesis so much favored previously that political rule originated in conquest—but derives primarily from the idea of earthly representation of the divine power that stands at the world's origin.[4] Thus, if the construction of cultures is nourished by religious roots, it is astonishing that cultural anthropologists scarcely ever define the concept of culture in terms of the religious roots of cultural development. If one implicitly or, like the Berlin philosopher Michael Landmann, explicitly regards the human being as "creator" of culture, one must deal with the difficulty that people of earlier cultures regarded their cultural way of life "as a divine gift or natural dowry," as Landmann himself says, and not as a human product.[5] Can that be set aside as a mere error? Is it enough to say, with Landmann, that "the creative force, however it was objectively already active, had not yet been subjectively discovered?" There may be a partial truth there. But it would only be legitimate to simply replace the self-understanding of ancient cultures with the creative productivity of the human spirit if such a self-understanding could be derived from that presupposition. However, the proof has not yet been provided by any theory of the sort.

3. The difficulty is even clearer if we combine the question of the origins of culture with the matter to be mentioned here in third place, namely, the origins of language. This is the most deeply disputed subject. While the great antiquity of burials and the resulting conclusions about the antiquity of religion are not in doubt and the fundamental significance of religion in ancient cultures is scarcely so, the matter of language is different. Most authors presume, more or less as a matter of course, that the construction of language originated either in dealing with matters of perception or in the need for social communication

about what was perceived. But there are indications that other factors were also in play at the beginnings of language. We are interested primarily in the religious factor.

These indications are found, first of all, in the realm of the psychology of infants' acquisition of language. Jean Piaget has demonstrated that the acquisition of language is closely related to children's play.[6] But play, human play at any rate, is in turn not the kind of innocent phenomenon Piaget thought it was. Rather, in its origins as a *representative* and *imitative* phenomenon, it is bound up, from a cultural-historical point of view, with the sphere of the cult. If children's acquisition of language is associated with children's symbolic play, in which toys serve to make concrete what is present in memory and imagination, then the transition to naming by means of language is understandable. For such play addresses its object in ecstatic fashion; thus, a stick makes it possible for the child to ride a horse, and it is only the riding that constitutes the true content and object of the play. But if children's language arises in connection with such ecstatic possession, it is probably no accident that Piaget speaks of the "mythic" and "animistic" features of children's developing language, even at the age of four to seven years. Without noticing it, he is here close to Ernst Cassirer, who, in his "philosophy of symbolic forms," sought the origins of language in mythical words: The object itself appears in the vocal sound. The word is not our subjective designation of the object but the manifestation of the object itself (compare the Hebrew *dabar*). According to Cassirer, it is only the connection to the mythical word in which the object itself is present that explains the function of language in representing objects and facts.

Likewise, in the question of the origins of language within human history, the real puzzle is its representative function—the representation of objects in words and sentences. Animals understand *signals* and give signals to each other aimed at influencing the partner's behavior. But the decisive step in language building is the transition from *signals* to *namings* directed at objects. How should we understand this developmental step? It is often assumed—and this seems natural in view of the vocal sounds that accompany children's play—that the *imperative* is the original form of both words and sentences. From the imperative comes *naming,* which originally was still close to invocation. But it is only the ecstatic surrender of play to its object that makes it understandable that words, as names, are connected to objects, and the activity indicated by the word thus becomes the activity of what is named, the now "objective" object itself, so that noun and verb are separated. The prehistoric transition from signal to naming has also been connected to play, specifically with the festive games that originate in the cult. At this point, I must quote from my book *Anthropology in Theological Perspective*:

The emotion [*Ergriffenheit,* seizure] of play, in fact, best explains [how] the sound replaces the object being played with, evokes the associated meaning, and represents for the individual the objectivity of the object. The presence in the word of the absent object constitutes "the essence of the word" as symbol. It appears originally as a mythic word, for in the mythic word the object is not only present, but acts as present.[7]

That the sounds accompanying the activity of play appear as the activity of the object made present by the word: that is

mythical causality, and it is at the same time the form of the verbal statement that attributes the activity experienced to the object named. The emotional involvement of the player with his or her object makes both the one and the other comprehensible. This means that language arises out of an originally religious emotion. It is thus no less a human creation, but precisely as such it, like all creative human activity, is indebted to experience and inspiration.[8]

I have spent so much time on the example of language because language is not only fundamental for the phenomenon of culture but also has rightly claimed a fundamental philosophical significance in current philosophical discussions. If the constitutive significance of the religious thematic for language is illuminating, that is important evidence strengthening the thesis that human life is ultimately sustained and kept in motion out of a deeper religious stratum.

Certainly what has just been said still leaves open the possibility of saying, with Durkheim and Habermas, that religion was foundational for the beginnings of human history and the construction of cultures, and even for the beginnings of human language (as Habermas also agrees), but that modern human beings have emancipated themselves from these religious beginnings and taken charge of their own lives. It must be admitted that something of the sort is not unthinkable and that the fact of modern secular culture, at least at first glance, affirms that such a divorce from religious foundations rooted in natural history is really possible. Still, we must ask whether a complete abandonment of the subject of religion should not be judged to be a repression phenomenon that, like every repression of something essential to the human being, will have negative consequences. Sigmund Freud investigated the consequences for the human situation and the individual human life of repressing basic sexual urges. Might not the repression of the religious dimension of human life have similarly destructive consequences?

There are indications that may confirm this suspicion. First, there are the feeling of alienation and lack of meaning that seem to be spreading in the world of secular culture. The two are closely connected. Alienation means, in the first place, that human beings cannot integrate certain parts of their daily lives into their sense of identity; they remain alien, and these very aspects then appear meaningless as well. But a human being needs a meaningful orientation to his or her surroundings, a meaning not given to things by the person himself or herself and consequently one that could not be arbitrarily constructed in a quite different way. Precisely this arbitrariness and transferability of meanings are expressions of the fact that the modern, secular world has no meaning in and of itself that trumps human caprice as a measure of human behavior and finds expression in the order of the world. The result may well be a feeling that life is utterly meaningless. Human beings can no longer find their own place in the world, their own identity, their sense of self. Without a meaning that sustains their lives, they feel alienated from themselves. The Vienna psychologist Viktor E. Frankl repeatedly referred to this experience of a "deficiency of self" and its effects. He correctly put his finger on a sickness of the age arising from our secular world's having forgotten God. In his view, it is ultimately responsible for the rapid increase in neurotic illnesses and especially the increasing numbers of suicides.

I see the second indication that the emancipation of the secular world from its cultural-historical religious roots has not led to a stable condition of things in the declining legitimacy of social institutions, and especially the political order of society. In the spheres of the family, of law, and of education, the solidity of an order of life prior to the individual is vanishing. It is not only that the traditional form of these institutions appears to be changing. Rather, the justification for orderings of common life beyond the individual is no longer persuasive. The demand that individuals should accommodate themselves to such existing orders of things is regarded as unreasonable, even when people, in fact, continue to live within such an order. This is true especially of the political order. The rule of some people over others is bearable only as long as it rests on an order that is commonly believed to be beyond all human caprice and manipulation, an order that is considered to be superior also to the arbitrary will of those in power and against which their behavior must be measured. The order of laws, and ultimately the constitution, comprises a first such measure. But when the order of laws itself is only something made by human beings and can be altered by relatively coincidental majorities, the claim of the order of law on the obedience of the citizens is itself in doubt. The modern state's divorce from religion as the measure of its order leads, as a consequence (and as Ulrich Matz has rightly seen) to lack of belief in the legitimacy of political governance of any sort. Thus, the legitimacy of the secular state is lapsing, but the same is true of other institutions of secular society as well. The insidious undermining of the authority of these institutions in the consciousness of the citizens will certainly not lead to their immediate collapse. The institutions continue to function more or less well. The average citizen is not too badly off, and as long as that is the case, there is no acute danger that the social system will fall apart. But such a system is vulnerable in times of unusual pressure, whether it comes from within—for example, through a sustained decline in the economic situation—or from without. The creeping crisis of legitimacy of the secular state and its order of laws may for the moment be nothing more than a tiny cloud on the horizon. Nevertheless, it is a questionable assumption that a social order completely loosed from its religious roots can survive in the long term.

Some decades ago the consciousness of secular culture was shaped by the expectation, as formulated by Max Weber, of the fateful advance of the process of secularization and that religion would necessarily be pushed farther and farther toward the margins. Peter L. Berger still presumed this in his 1969 book *A Rumor of Angels*.[9] But only a few years later, in 1973, Berger wrote another book, this time titled *The Homeless Mind*.[10] This book showed that the process of secularization or—as Berger preferred to say—the "modernizing" of society through efficient organization, bureaucratization, and industrialization was by no means proceeding in linear fashion. The process could not continue forever without creating counter-forces. In other words, the modernization of society leaves all the elements of life's meaning to the caprice of private choice—not only religion but also art and all the content of cultural tradition. Public life is released from all connections to obligatory bases of meaning. The arbitrariness of what has meaning destroys the very sense of meaning, which seeks something that is beyond human caprice and therefore binding, something to which one can hold fast. Since the public order of life no longer reveals any such normative meaning, according to Berger, the result is that everywhere in secular societies there arise counter-movements that are subcultural in nature. What is important about these is not so much their external expressions: These can be disruptive or can take the form of silent absence or of political protest. What is common to these counter-cultural movements, according to Berger, especially among young people, is their resistance to the system of modern society and its restrictions, all of it seen as empty and meaningless. The roots of such resistance are ultimately religious in nature, for that a human being cannot live without normative meaning ultimately signifies that he or she cannot live without religion. Hence, today we no longer expect that religion will be increasingly replaced by secular culture. The question is no longer how much longer religion may continue to exist or when religion will entirely die out. The question today is rather how long a secular society can survive, having loosed itself from its religious roots. Religion, therefore, will not disappear. For, at any rate, the world religions themselves have continually experienced the decline and collapse of the ruling systems of this world.

The conclusion of these reflections is that the religious beginnings of humanity may well not have belonged merely to an initial phase of human history that we have now overcome. The notion that religion may have had constitutive significance for the beginnings of human culture and language but has now been replaced by a self-emancipating, purely secular form of human life—this notion still needs to provide some proof of its assumptions.

In the secular world as well, the dimension of an emotion that can only be called religious has from the beginning been at least implicitly present and effective in the life history of every human being. When an adequate language is not found for this dimension, the result is some form of ossification in spiritual development. After all, the human being is not from the very beginning an "I" that remains identical with itself through all the shifts of experience, the unified and unifying subject of all its behaviors, activities, and experiences. This "I," as idealistic philosophy of the subject conceived it, is a fairly late product of individual development. As we know, children learn to use the word "I" fairly late, after they have already learned to use their own names in relation to their own bodies. The findings of contemporary psychology of development (for example, the researches of Jane Loevinger) converge with observations drawn from analysis of the acquisition and use of language, leading to the conclusion that the "I" does not exist much before the use of the word is learned. At any rate, at the beginning, there is a "symbiotic sphere" (a period of common life) in which the child does not yet differentiate itself from the mother but is bound up with her in a single sphere of life. And it is only after the child has learned to distinguish the objects in its world that it is able to comprehend, in relating to those things, that it is itself separate. In this experience, the expectations and demands of the other play an important part, as the social psychologist George Herbert Mead demonstrated. But even when the word "I" begins to be used, there is still not, and will not be for a long time, a "solid and enduring self" in the sense of the idealists' concept of the subject. At first, the "I" is not yet protected against becoming something else at any moment. The stability of the self is the result of a tedious process of identity construction. This process can in large part be described, in terms of social psychology, as an internalization of judgments, expectations, and norms from the social environment. But something else is also required for what Erik Erikson called "healthy personality development," namely, what Erikson called "basic trust." This refers to a child's openness to the world in which it is growing up, an expectation of sustenance, nourishment, and support. "Expectation" says too much, however, because there is not yet a subject that could expect something or do acts of trusting. Rather, the child, in its "symbiotic" union with the mother, exists in a very fundamental way outside itself. The "ecstasy" or emotion comes before the existence within oneself, the "hypostasis," the person. This is continued later in the ecstatic emotion of play and in the initial use of language; even for adults, all their spiritually creative moments (and erotic life as well) take the form of ecstatic emotional possession. What is it that seizes us then? What, exactly, is the sphere of life at the beginnings of infant development that developmental psychology calls the symbiotic sphere that binds the child to its mother? Is it society, first represented by the mother? Or is it not an undetermined whole that from the beginning exceeds all limitations because it precedes all limitations and only manifests itself in particular things and appearances that are gradually distinguished as such? At the beginnings of infant development, the mother represents the world to the child, and not only the world, but God as well, the ultimate sheltering and sustaining horizon for the child's life, the horizon in relation to which the inchoate ecstasy of the symbiotic life of infancy's first weeks is transformed into that amazing trust that opens itself to a world we grownups know only too well is so little deserving of our trust. Even the mother cannot justify such overwhelming trust in the long run; she cannot fully meet its expectations. The more experience shows that even our parents can only protect and provide security to the child in a limited way, the more does it require a religious upbringing in order to retain that limitless trust that is so important for healthy personal development. Religious education transfers the limitless trust of the child in its mother, and then both parents, to the object that is always the only adequate one to receive such unlimited trust: God. The mother only represents God at the beginnings of infant development. The trust bestowed on her is always infinitely greater than her limited strengths and her inevitably also limited devotion to the child. To that extent, the symbiotic ecstasy of human beginnings and the child's so touchingly limitless trust find only in God their real ground and anchor. And human emotional life, in which the whole of our life is always present to us in some coloration and mood, requires even in later years the expansiveness that only religion holds open to it, beyond family, beyond society, beyond this world.

Is this leading to something like a demonstration of the existence of God? I think not. What, in fact, results from the arguments presented is only this: The human being is by nature religious. Religion, whatever form it may take, is a necessary dimension of human life, and if it dies, one must expect consequential deformations in the development potential for human lives. But the fact that religion is a constant of human reality is no guarantee that God exists. It might be that the human being, in its nature incurably religious, is a failed product of evolution, at least in this regard, that is, to the extent that it has fallen victim to a natural illusion. The existence of divine reality can only be proved by God, in that God's reality touches human beings so that the world and their own lives become recognizable as God's creation. God's being opens itself only to religious experience, which, of course, is also expanded by thought. But the orientation to religion (and perhaps religious awareness itself) precedes every particular and special religious experience. It belongs—however suppressed it may be—to the "nature" of the human being. This does not resolve the confrontation with atheism. But the basis for engaging the atheistic arguments of Ludwig Feuerbach and his successors is changed: The human being is not first of all one that can be described in purely secular terms and that then secondarily, for reasons that require special clarification, somehow came to the strange notion of projecting ideas about God into an imaginary heaven. Instead, the human being is from the beginning, by nature, religious, and the secular form of life is a late product and an exceptional form of human culture. What can then remain in dispute is only whether religion is an illusion that is unavoidable (because it is given in human nature), or whether the incurably religious nature of human beings is the seal of their origin in their Creator. Christian faith teaches us the latter. The ground for recognizing that God is real can come only from God, and this happens when God reveals himself to us. But that God is self-revealed to us would have to remain a message foreign to us, an assertion that does not reach us, if we did not thereby come to the awareness that we, in the depths of our own human reality, already belong to God as his creatures.

# → 8 ←

# Human Life

## *Creation Versus Evolution?*

E VER SINCE its first publication by Charles Darwin in 1859, the
doctrine of evolution of living forms and species by natural selection among individual variations within a given population in a struggle for survival has been a matter of dispute among scientists, and it has
become an ideological controversy. The dispute among scientists, however, did not center on the issue of whether there is or can be a process of evolution of higher organized species from lower forms of life.
Rather, the scientific discussions were mainly concerned with the question of whether the principle of natural selection is sufficient to explain
the process of the emergence of ever new and more complex forms of
life.

There are a number of difficult questions related to this issue. First
of all, what is the standard requirement according to which selection
operates? Is adaptation to external conditions the standard of fitness
for natural selection, as the mechanistic interpretation of Darwinism in
the late nineteenth century assumed, or does the spontaneous productivity of genetic variation lead to the discovery of new natural "niches"
for survival and consequently of new objects for adaptation? Furthermore, can a continuous and cumulative occurrence of small variants
under the pressures of natural selection issue in the emergence of a new
species, or do small changes tend to disappear because they don't fit in

the overall system of the organism and of its functioning? Would, then, a "fulguration" of a complete new scheme of organization be required for a new species to emerge? Finally, how is the apparent direction of the evolutionary process toward ever more complex forms of organic life to be accounted for? These are but a few of the more important riddles that have plagued Darwinism from the start and still continue to vex its defenders. Nevertheless, the general perspective of the Darwinian theory has been victorious, even though it is still hypothetical and the evidence for it rests on a somewhat defective fossil record rather than on experiential demonstration. For all its difficulties, the theory of evolution still provides the most plausible interpretation of what is known about the history of organic life on this planet.

The resistance against the new theory from the side of the churches had been predictable, since it stood in clear contrast, if not contradiction, to the traditional concept of Creation. For many centuries, it had been taken for granted that, as according to the biblical account in the first chapter of Genesis, the species of plants and animals had been created by God on the fifth and the sixth day of Creation and have remained unchanged ever since. Even among those, however, who do not cling to biblical literalism, it seems unacceptable that the theory of evolution could replace God's purposive action in bringing about the different forms of life by a mechanical process of nature. In this controversy, the point is that, before Darwin, the purposive action of the creator had been understood to provide the only explanation for the fact of different species of animal life. Therefore, the proposal of a natural explanation of the same result was taken as a denial of God's purposive action in the creation of living forms. In principle, of course, the assumption of God's purposive action need not have excluded the use of natural causes in the execution of the divine purpose. In historical fact, however, in the situation after Darwin's book *On the Origin of Species* had been published, explanations by divine purposes and by the mechanical operation of natural causes were taken as alternatives.

## ATTEMPTS AT THEISTIC EVOLUTION

Given the antagonistic climate of the early discussions of Darwin's theory, it is astonishing that, from their beginning, some leading British churchmen and theologians tried to reinterpret Christian doctrine in the light of the perspective of evolution. The most remarkable of these attempts was a book edited by Charles Gore in 1889 under the title *Lux Mundi: A Series of Studies in the Religion of the Incarnation*. As the title suggests, the book reinterpreted the incarnation of the divine *Logos* in Jesus Christ in terms of providing the culmination of the evolution of life. While the process of natural evolution culminates in the emergence of the human race, so the history of the human race reached its climax in the Incarnation.

To a certain extent, such a theological scheme was suggested by early Church Fathers like Irenaeus. But now in post-Darwinian times, the picture of a salvific history of humanity leading toward the event of the Incarnation was being immensely broadened by including the process of natural evolution of life as a prehistory of that history of salvation.

Interestingly, the authors contributing to *Lux Mundi* did not take Darwinian evolution to describe a mechanical process but rather a historical process. That was hardly warranted by the situation in the development of evolutionary theory around 1890. *Lux Mundi* rather pointed beyond that situation to a future concept of "emergent" or "organic" evolution, as it was proposed in 1923 by Lloyd Morgan: Emergence means that, at each step of the evolutionary process, something new comes into existence. It does not merely "result" by mechanical necessity from past conditions. This concept of emergent evolution vindicated the positive evaluation of Darwinism by the group of *Lux Mundi*, who had celebrated the new theory for doing away with the God of deism who had been responsible for the beginnings only, while now God could be seen to be active in every new turn of the evolutionary process. The concept of emergent evolution overcame the mechanistic, reductionistic way of describing Darwin's theory, and the tendency to emphasize the element of the new in the sequence of evolving forms of life was further strengthened by the realization that major steps in the evolutionary process need "fulgurations" of new schemes of organization rather than a sequence of small steps of cumulative variations.

## EVOLUTION AND THE BIBLICAL WITNESS

After providing the stage for a theological discussion of evolution, I now turn to the crucial issue of whether a theological appropriation of the doctrine of evolution can do justice to the biblical witness on the creation of animal species by God. In a subsequent section of this chapter, the same question will be asked with regard to the human race. What has been said so far on the further development and refinement of the theory of evolution since Darwin will prove helpful in the attempt at answering both of these questions.

When we turn to the biblical witness on Creation, the first thing must be to remind ourselves of the fact that the biblical texts are historical documents and have to be interpreted accordingly in terms of what they were trying to say at the time of their composition. This principle of the historical interpretation of the Bible is the core issue in all discussions with creationists. Historical interpretation reads the biblical affirmations relative to the context of their writing, to the concerns of their authors at the time of their writing, to the knowledge they had at their disposal. Such historical interpretation does not imply that the biblical affirmations, being limited to their own time, have nothing to tell readers of a much later period. But whatever they have to tell us, they convey it precisely through their historical particularity. To the degree that their affirmations have universal significance, it is inherent in their particularity. Otherwise, it would not be the meaning of the biblical affirmations but a meaning the modern interpreter reads into them. Furthermore, the historical reading of the biblical affirmations does not preclude their appreciation as the Word of God, the Word that addresses us as it does every generation of human persons. The Word of God expressed in the biblical affirmations is, however, a unified entity. It is the Word of God that became incarnate in Jesus Christ. To read or hear the Bible as the Word of God is to relate each particular biblical affirmation to the whole of the biblical witness and to interpret the detailed, historically distinctive affirmations in that light. Therefore, reverence for the Bible as the Word of God does not stand in opposition to a careful historical scrutiny of each individual sentence.

With regard to the biblical report on the Creation of the world in the first chapter of Genesis, this means that we have to read its affirmations as witnessing to the God of Israel who is the creator of the world by using the natural science of the sixth century before Christ, that is, Babylonian wisdom, in order to account for the sequence of creatures coming forth from God's creative activity. The relevance of this report in our present situation, then, is primarily the encouragement to use the science of our era in a similar way for the purpose of witnessing today to the God of the Bible as creator of the universe as we know it. This is the authority of the biblical report on the Creation of the world. It calls us to try our own theology of nature but, in doing so, to remain true to the peculiar and distinctive nature of the God of Israel, just as the authors of the priestly report on the creation of the world did in their own era.

The authority of the biblical report does not require us to consider every detail as the last word on the respective issue. Many statements are inevitably indebted to the limited knowledge about nature in the sixth century BCE. One example is the idea that the experience of rain is evidence of a huge supply of water in heaven above the clouds, comparable to the oceans on earth. On this assumption, it is astonishing that the waters above the clouds normally remain separated from those beneath. This is explained by the idea (Gen. 1:6ff.) that God created a vault to keep the waters above from pouring down. This mechanism is completely rational, and yet this beautiful and important detail can no longer be part of our conception of nature.

The same applies to the assumption that all the different types of creatures, and especially all the different species of plants and animals, were created in the beginning and remain permanently unchanged. This idea is an example of the mythical attitude of mind in early cultures, where generally, as Mircea Eliade told us, the world order was conceived as having been built in the "original time," *in illo tempore,* without later change. By contrast, the modern knowledge of nature provides sufficient evidence for assuming that the natural world is in a continuous process of becoming. This means that there is a continuous emergence of new types of creatures, along with the disappearance of others. All this belongs to the picture of nature with which we work.

## CONTINGENCY AND NEWNESS
## IN NATURAL HISTORY

Does the modern picture of nature in terms of continuous change contradict the biblical doctrine of Creation? It is certainly at variance with the image in the first chapter of Genesis that the whole order of creation was produced in six days and continues to exist unchanged. But within the Bible as a whole, we find other pictures of God's creative activity. In the prophetic writings, for example, we learn that God is continuously active in the course of history and that once in a while he creates something quite new (Isa. 48:6ff.). That is not to deny the creation of heaven and earth in the beginning. Yet Second Isaiah (the author of Isa. 40–55) takes that as an example of God's continuously creative activity. This, then, is the model of a continuous creation that is coextensive with the course of the world's history. In this model, the creation of heaven and earth is much closer to the modern understanding of nature in terms of a history of the universe than is the image of the six-day creation in Genesis. Such a conception of continuous creation does not have difficulties with a doctrine of evolution, according to which the different species of animals emerge successively in the long process of life's history on earth.

There is one requirement, however, which must be met if the concept of evolution is to be compatible with a theology of nature based on the biblical idea of God. This is the acknowledgment that something new occurs in each and every single event. Newness also occurs in the emergence of new forms of life in the process of evolution. This element of newness or contingency was not in the focus of the early mechanistic interpretation of Darwinism. However, it has been increasingly emphasized in the conception of epigenesis, which means the emergence of something new. Contingent newness belongs in the concept of *emergent evolution.*

Why is the element of contingency so important in a theological appropriation of the theory of evolution? The reason is that the Bible conceives of God's relationship to the world in terms of free creative acts in the course of history as well as at the beginning of this world. In the first chapter of the Bible, this concern for God's freedom in his creative activity is expressed in the concept of the divine word, which brings about its effect in the most effortless way. In each creative act, God's freedom brings forth something new simply by the word. Therefore, the history of the world is seen as an irreversible sequence of contingent events, notwithstanding all the regularities that can be observed in its course. Consequently, a concept of evolution in terms of a purely mechanical process would not be easy to reconcile with the biblical idea of God's creative activity; yet the concept of an epigenetic process of evolution with something new to occur in virtually every single event is perfectly compatible with it.

In addition, God's creative activity does not exclude the employment of secondary causes in bringing about creatures. In the priestly document on Creation from the sixth century BCE, preserved in the first chapter of Genesis, the Creator calls on the earth to bring forth vegetation (Gen. 1:11). And again it is the earth that is called upon to produce animals, especially mammals (Gen. 1:24). If our creationist friends today would adhere, in this case, to the letter of the Bible, they could have no objection to the emergence of organisms from inorganic matter or to the descent of the higher animals from those initial stages of life. In the biblical view, such a natural mediation does not contradict the affirmation that the creatures are the work of God. For, in the next verse, it is explicitly said that *God* made the beasts and the cattle and everything that creeps upon the ground (1:25).

Of course, the biblical text does not tell anything about the higher species of animals as having evolved from lower ones. But that is an issue of secondary importance, if compared with the question of whether the act of creation must be conceived of as an immediate action of God without any mediation by other creatures. This question, however, has been answered already. The immediacy of God's creative action with reference to its creatures is not impaired by secondary causes, since their activity is not on the same level with that of the Creator.

## THE APPEARANCE OF THE HUMAN SOUL

The case of the human being is a special one because human persons are related to God in a special way. This fact is indicated by the importance of religion of one form or another in the history of the human race. Human self-consciousness seems closely connected with some form of awareness of the Divine. In the Bible, this close relationship to the origin of the universe is expressed in the idea that the human person has been created in the image of God. Therefore, the human being represents the Creator himself with regard to the rest of God's creation. Does that not require that the human being was created by God alone, without the cooperation of other, earlier creatures? In the first chapter of the Bible, no such cooperation is mentioned. Does that mean it is excluded?

The older report on the creation of human beings in the second chapter of Genesis does not justify such a suggestion because it says that the human body was formed of "dust from the ground" (Gen. 2:7). That seems to be roughly equivalent to the role of the earth in the first chapter of Genesis, when God addresses the earth to bring forth plants and animals. Therefore, our body is perishable, which is to say, it will return to the earth. Only the human spirit is said to come directly from God, as the second chapter of the Bible describes it: God breathes his breath into the figure he formed from the dust; he "breathed into his nostrils the breath of life" (Gen. 2:7). Correspondingly, with our last breath, we return the gift of the spirit to God, as the psalm says, which, according to the Gospel of Luke, Jesus quoted when he died on his cross: "Into thy hand I commit my spirit" (Ps. 31:5, Luke 23:46). In the moment of death, the spirit or breath is separated from the body, and as Ecclesiastes says, "the dust returns to the earth, as it was, and the spirit returns to God who gave it" (Eccles. 12:7).

Does that mean that we are allowed to think of the human *body* as coming from the process of evolution of animal life, but not so of the human soul and spirit? This could seem to be required by the older creation story when it says that the Creator breathes the breath of life into the figure formed from clay and that thereby the human creature became a living being (Gen. 2:7). The Hebrew term here is *nephesh hajah,* and *nephesh* was often translated as "soul." Thus, God is presented here as creating the human soul by breathing the spirit of life into the nostrils of the human body. It was from this sentence that the old Christian creationism of the Patristic period derived its theory about the origin of the human soul: While the body of each new individual was considered to come from the chain of propagation, each individual soul was believed to be added to the body by the Creator himself. But this Patristic creationism presupposed the independent status of the soul as compared to the body, an idea that is in keeping with Platonism but not with the Hebrew Scriptures. In the Old Testament, *nephesh hajah,* which we translated by the term *soul,* is not independent from the body but the principle of its life, though it is not the origin of life itself. The *nephesh* is only the continuous hunger and thirst for life. The root meaning of the word is "throat." It is in constant need of the spirit of God, that productive breath or wind that animates the soul and, through the soul, its body. It is only through the spirit that the human being becomes a "living soul," as the phrase in the creation story goes.

To be a "living soul," however, is not a distinctive prerogative of the human being. According to the creation story in the first chapter of Genesis, the "breath of life" is in all the animals, the beasts on the ground, and the birds in the air (Gen. 1:30). This corresponds exactly to the idea in the earlier report on the creation of the human race, where God breathes the breath of life into the figure of clay so that it comes alive. If the animals have the breath of life within themselves, although they are products of the earth that was summoned by the Creator to bring them forth, then there is no difference from the creation of the human being with regard to its description as "living soul," *nephesh hajah.* The difference of the human being from the other animals is not that the human being has a "living soul," but that it is destined to exist in a particular relationship to God, so that it is called to represent the Creator himself with regard to the animal world and even to the earth (Gen. 1:26).

The excursion into biblical exegesis was necessary to meet the charge of modern creationists that the doctrine of evolution and especially the derivation of the emergence of the human race from the evolutionary process of animal life contradict the biblical creation stories. When, in the Bible, animal life is seen as a product of the earth and the formation of human life as "living soul" is understood as analogous to animal life, then there is no reason that the human being should not have emerged from the evolution of animal life. The idea of evolution as such is a modern concept and cannot be derived from biblical conceptions.

But it is not opposed to the basic concerns of the biblical conceptions of the origin of animal and of human life. This can be affirmed as long as the modern idea of evolution does not exclude the creative divine activity within the entire process of evolution.

## BEYOND MECHANISM TO EMERGENCE

The doctrine of evolution is open to a theological interpretation when it is not conceived in terms of a mechanical process, based on the principle of natural selection, but as describing a process of emergence, in the course of which the productivity of life continuously produces something new. The element of contingency in this concept of emergent evolution secures its openness to the creative activity of God in this process. That each form of life can be understood as a creature of God is not dependent on the idea of purpose, the assumption of a purposeful adaptation of each species to the conditions of its survival in its environment. In earlier times, it was assumed that such purposeful adaptation presupposes and demonstrates the intelligent will of the Creator and is not reducible to other causes.

It was this assumption that Darwin destroyed by explaining the adaptation of a species to its environment as a result of natural selection. But the principle of natural selection does not exclude the continuous activity of the Creator in the very productivity of life. The superabundant creativity of life and the creative action of God are not alternative notions—no more so than is the productivity of the earth, which is called upon by God in the biblical creation story to bring forth vegetation and even animals. The spontaneous creativity of life is the form of God's creative activity.

In a modern perspective, self-organization is characteristic of life on all levels of its evolution. It accounts for the spontaneity in all forms of life, and it is in the principle of spontaneous self-organization that we have to perceive the roots of human subjectivity. Self-organization is the principle of freedom and of superabundance in the creative advance of the evolutionary process. Human self-consciousness is its highest manifestation as far as we can see, as it allows us to integrate all other consciousness into the unity of our individual self. Self-consciousness itself is not merely a given fact of nature, however. In each individual life history, it arises from the early stages of the development of our consciousness. Self-consciousness itself is a product already of the creativity of life within each one of us, a product of the creative activity of the divine Spirit. The creative self-organization of life in the process of evolution since the transition from inorganic matter to the first organisms corresponds to the blowing of the divine wind, the Spirit of God that breathes life into ever new creatures and thus blows through the evolution of life until it overcomes all perishability in the resurrection of Jesus Christ. The death of individuals is due, according to the biblical witness, to their limited share in the divine Spirit (Gen. 6:3). To Jesus, however, though a finite being himself, the Spirit of life was given "without measure" (John 3:34). Therefore, he was raised from the dead by the power of the Spirit and transformed into a spiritual body (1 Cor. 15:44ff.), which is to say, into imperishable life, which is imperishable because of its unbroken participation in the divine Spirit who is the source of all life.

A Christian account for the evolution of life as expression of the divine Spirit blowing through creation cannot avoid some reference to the eschatological resurrection of the dead, the climax of the creative activity of the divine Spirit that was first realized in the resurrection of Jesus but is meant to embrace humankind in general by communion with Jesus and even, according to Paul, the world of other creatures, because "the creation itself will be set free from its bondage of decay and obtain the glorious liberty of the children of God" (Rom. 8:21).

To the modern mind, the biblical description of life as created by the dynamic activity of the divine Spirit may appear as merely metaphorical. Such an appraisal becomes even more suggestive when one realizes that the Hebrew notion of spirit means breath or wind rather than intellect. The images of breath and wind seem to be just images, without providing rational explanation.

The people of ancient Israel, however, took breath and wind quite literally to be the cause of life. This seemed to be confirmed by everyday experience: Life begins when a baby starts to breathe, and it ends with a person's last gasp. Modern people may no longer consider this intuitive evidence to provide a sufficient explanation for the beginning and end of human life. Yet breath is still more than an arbitrary image because it indicates the dependence of life upon a flow of energy that enters our body from the outside and passes through ourselves. Like a flame, our life is a process of exploiting a flow of energy by degrading the high-level potential energy of our food and the oxygen we breathe into a state of increased entropy. Life is an autocatalytic process of self-organization that exploits the gradient of energy in our environment, as the flame does by keeping its equilibrium at the price of slowly consuming a candle. The description of life by the phenomenon of breath that goes through us is an example, as is the flame, of taking advantage of a flow of energy by letting it pass through ourselves. This consideration is not specific to *human* life, of course, since it applies to all organic life even in its most primitive forms. But the wonder of life is in the abundance of increasingly complex forms that the principle of self-organization produces as it happens in the process of evolution. The human being is a very complex and specific example of such creative self-organization, in the development of an individual as well as in human culture. In an elementary way, self-organization takes place in the development of each individual, and there we can try to perceive a peculiar form of the work of the Spirit in human beings. This work of the Spirit is not yet the specific function of sanctification, but first the invigorating action of the Spirit in our personal development.

Human life is endowed with consciousness, memory, and self-consciousness, but all these characteristics have to develop in the course of our individual life. In the beginning, there is no self-consciousness, but only a disposition for its development. Even the acquaintance with our environment is a task that we can meet only through our own productive activity. Perception and consciousness of objects are things we share with the animals before us, but only to a modest extent do we react instinctively, as they do, to stimuli from the outside. Rather, we have to develop our own survey of our world in order to relate to the objects of experience, and we achieve that by the development of language. Language is a form of organization, an active organization of our world for ourselves. Though each individual person takes over language from the social context, the process of acquiring language is still a creative process of self-organization, as is the later appropriation of our cultural heritage. Yet such self-creative activity is not something we "make"; rather it is produced by some sort of inspiration that activates us. It is only in the course of that process that we learn to distinguish our own body from the objects around, learn to give our own body its name like other objects, and finally learn the use of the difficult word "I" and related words like "mine" or "my." Self-consciousness develops along this line. It is not there and complete from the outset but depends on the development of language, though later it becomes the center of our personal life. The key of our human life is language. Nothing is independent of it. Even the use and development of tools beyond primitive stages depend on language.

The world of language is not the only but the most distinctively human dimension of the activity of the Spirit in our lives. Therefore, we are not mistaken to use the word *spirit* in a special connection with consciousness and language. It activates us so that we actively develop our conscious life and language, but we do not "make" spirit in the specifically technical sense of that word. We are active in producing it, while we participate in our social context and in the language of our culture. Through and beyond that, we participate in a spiritual dynamic that does not originate with us and also surpasses and comprises our society and world. A sense of the surpassing mystery of life belongs to the human condition, precisely because we organize the objects of experience with the help of language in encompassing totalities that are themselves transcended by the surpassing mystery. Wherever humans encountered that mystery, they usually called it by the name of God. The awareness of the religious dimension of life belongs very closely to the specifically human form of consciousness and self-consciousness. It belongs to the origins of language. In the Bible, it is God who brings the animals to Adam "to see what he would call them" (Gen. 2:19). It serves the poetic purpose of that story that this episode is dealt with before the creation of woman, but it seems, rather, that language is a social phenomenon, not a solitary product of solitary individuals, and furthermore it may have had ritual origins in its beginnings.

Our secular culture tends to underestimate the encompassing importance of religion in the early history of human culture. This also applies with regard to the period of transition from prehuman animal life to human life in the full sense of the word. Burials belong to the oldest documents of human life, and their occurrence since the Paleolithic period, even before the early traces of visual art, serves as a criterion for when and where the transition from animal behavior to human culture has come to its conclusion, as argued by Karl J. Narr and Anthony F. C. Wallace. Thus, religion is constitutive for the beginnings of humankind.

Whether this final step—from biological to cultural evolution—took place only once is a matter of secondary importance. In the Bible, of course, all humans were understood to have come from a single pair of parents, Adam and Eve. But in the biblical creation story, this is not a matter of special theological emphasis but follows from the narrative's way of treating the creation of humankind in the form of one paradigmatic individual, Adam. In the Roman Catholic doctrine, the descendance of all human beings from the one Adam is still considered important because of the doctrine of original sin as inherited from Adam. If, however, the story of the Fall is to be read as a paradigmatic description of human behavior rather than in terms of a unique event in the beginnings of human history, then the Christian doctrine of original sin depends more on the affirmation that all human beings repeat the paradigmatic pattern of Adam's and Eve's behavior in the garden of Eden than on biological heritage. Therefore, the question of whether the transition from prehuman life to humanity took place in only one individual (or two individuals) or at several points within a larger group of individuals is a matter of secondary importance as compared to the spiritual nature of life in general and of human life in particular.

# ✦9✦

# Consciousness and Spirit

THE CONCEPT OF SPIRIT and propositions about the spiritual nature of consciousness have not enjoyed any special attention in current philosophical discussion. This may be primarily because we deliberately avoid unwelcome associations such as those that could remind us of traditional notions of spiritual substances alongside the material world and underlying the phenomena of physical behaviors. Nevertheless, there are good reasons to think that a discussion of the differences and connections between consciousness and spirit could be useful for a more precise investigation and clarification of the relationships between body and soul, something that appears for many to be a settled question but that has in recent years, somewhat unnoticed, again become a subject of serious discussion. An examination of the differences and connections between consciousness and spirit will also make it possible for theology to make its contribution to the dispute about body and soul and the question of the place of consciousness in the context of nature.

In modern philosophy's approach to the problem, René Descartes' doctrine of the existence of two substances, *res cogitans* and *res extensa*, has been to the present time the starting point for any discussion of the body-soul question. Karl Popper has shown that Descartes' theory of matter and the mechanistic movements of the body make it incomprehensible how the soul, which in his view is immaterial, could be able to move the body.[1] Popper has also shown how this difficulty in Descartes' system opened the way to the development of various theories of parallelism between body and soul, beginning with the occasionalists and continuing through Benedictus de Spinoza to Gottfried Wilhelm Leibniz. Cartesian dualism also gave the impulse for the attempts of the physicalists or materialists to demonstrate that proposing a second substance alongside the body was superfluous. This trend in philosophy was materially advanced by David Hume's critique of the notion of consciousness as substance. In the early days of English empiricism, John Locke was still able to assert that the idea of a spiritual soul, as a special spiritual substance, was demonstrated from the activities we can observe within ourselves, by the same evidence with which we construct the idea of bodies as the causes of the impressions on our senses.[2] Hume, in contrast, regarded the idea of a spiritual soul (mind) in the sense of a special substance different from the body as absolutely unintelligible. He affirmed that we receive no impressions at all from a spiritual soul that can be compared to the impressions on our senses that lead us to propose the existence of physical objects outside ourselves.[3] Subsequently, under the influence of physicalist and behaviorist methods of observation, the trend toward a reduction of the spiritual to functions and epiphenomena of physical processes continued to advance; and in recent decades, linguistic analysis has added another argument leading to the same conclusion, by reducing the idea of a subject within us to a mere mode of speech, namely, the word "I," which according to English linguistic-analytical philosophy has as its sole function to designate the particular speaker and hence is classified terminologically as an "index word." If one strictly limits the word "I" to that function, there is no reason to take it as an indication of the existence of a subject housed in the body in the sense of a spiritual soul. Gilbert Ryle, in his book on the concept of consciousness, joked about the "myth of the ghost in the machine."[4] That made it all the more surprising that a thinker of Karl Popper's range, whose thought was deeply rooted in the empiricist tradition of philosophy and scientific theory, publicly confessed his belief in the "ghost in the machine" and, together with the highly regarded neurologist John Eccles, developed a dualistic model of the relationship between body and soul, especially in regard to the relationships between human consciousness and the brain. The basis for this was Eccles' comprehensive neurological description of the human brain and its functions. The book Popper and Eccles published together, *The Self and Its Brain*, was an authoritative contribution to the reopening of the debate about the relationship between brain activity and consciousness.[5]

Popper's main argument against physicalist or materialist conceptions of the relationship between physical and spiritual phenomena is aimed at the fact that, at least in their more radical forms, these not only deny the reality of consciousness, but in particular they are not able to offer any adequate explanation of the possibility for technical ideas and other human cultural creations, at least not such that one could account for the specific logical structures and constructive forms of such creative ideas. The key point is that such ideas do not simply arise as the result of physical processes but are the product of human design. Popper called the world of such technical and other cultural ideas a "third world" alongside the worlds of physics and of human interiority, in order to give special emphasis to its underivability from the world of physical processes. However, according to Popper, a physicalist or behaviorist interpretation of human activity also fails because it is unable to explain the higher functions of human language, especially those of description and argumentation.[6] In particular, physicalism can clarify neither the concept of truth nor the truth-claims of propositional statements.

The positive basis for Popper's critique of physicalist or materialist interpretations of human reality is his interpretation of the evolution of life—and not only of life but of the universe as a whole—by developing the idea of a process of emergence.[7] Such a process is characterized by the fact that marginal deviations can sometimes, under selective pressures from the environment, inaugurate a new epoch of development. In this way, qualitatively new emergent qualities that "appear" in the course of evolution represent, in each case, something new and unpredictable. Popper follows Sir Alister Hardy's "organic theory of evolution," convinced that, in this process, not only changes in physical organization but also changes in the behavior of living things can become the natural selections of the process of development.[8] In this way, Popper achieves a general theoretical basis for his specifically anthropological thesis that, in the development of the human species, the acquisition of language—that is, the shaping of a new form of behavior—has become a factor in natural selection: "[T]he evolution of language can be explained, it seems, only if we assume that even a primitive language can be helpful in the struggle for life." This means for Popper that the phenomenon of "language, once created, exerted the selection pressure under which emerged the human brain and the consciousness of self."[9]

The appearance of self-awareness and a self-conscious subject is thus, in Popper's view, dependent on language and not vice versa. According to Popper, the same is true also of the development of the human individual: "[W]e are not born as selves. . . . [W]e have to learn that we are selves."[10] Of course, it is not only a question of language. The acquisition of language presupposes the discovery of the world through perception, especially the development of a sense of the constancy of objects, the stability of spatial relationships, and the identity of objects over time. All these steps of development must be completed before there can be human self-awareness. Therefore, Popper writes, "Temporally, the body is there before the mind. The mind is a later achievement . . . ."[11]

This view of the origins of self-awareness, in dependence on language and thus also on a sociocultural milieu, recalls the thought of George Herbert Mead, who also regarded the human self as the product of social interaction and especially of language. But Popper is far more radical than Mead had been. According to Popper's argument, the dependence of self on language is not only a matter of the idea we acquire of the self; Popper thinks that the human mind itself arises out of contacts with the social environment and especially the acquisition of language, although he acknowledges that there are natural, inherited predispositions for such a process. To me, these assertions about the process that leads to the emergence of self-awareness seem quite plausible. Of course, Popper's view also brings with it a whole host of problems. One of these is that a sharp distinction must be made between consciousness itself, in the sense of awareness of objects, and human self-consciousness, including the reorganization of the whole conscious life associated with its emergence. This distinction is indispensable to Popper's conception because language naturally presupposes at least sense perceptions and so can only be self-awareness, not consciousness itself, something that only develops in us as a result of the acquisition of language. In his dialogues with John Eccles, Popper in fact repeatedly emphasized the distinction between consciousness and self-awareness, and in doing so, he indicated that only self-awareness is to be regarded as a human privilege, while at least a momentary awareness associated with actions of perception must be attributed even to the higher animals.[12] The idea of a human mind as product of language must, therefore, be applied particularly to human subjectivity, characterized by self-awareness, and is to be distinguished from consciousness in a more general sense.

Another problem appears at certain points where Popper calls the world of culture and language, what he calls "World 3," a world of thought-content and the "products of the human mind."[13] What, in this context, is the status of language? If human self-consciousness only comes about through language, one can certainly imagine that it has a reverse effect on the further development and use of language, but language as such can then no longer be described simply as a product of self-conscious subjectivity; otherwise, the beginnings of self-awareness would be explained by a factor that, in turn, owes its origins to self-awareness. That would be a vicious circle. If human self-conscious subjectivity first comes about through language, then, in its origins, language must somehow precede self-awareness but also be different from the physical world since the differentiation of self-consciousness from the world of physical objects is to be its result. We may be permitted, in order to illustrate this, to use the concept of field to designate what we apparently must presuppose if we are to understand human self-consciousness as the product of language or, at any rate, as developing in close connection with language. The "field" within which the creation of language takes place may then be termed a "spiritual" field. Such a designation cannot be deemed inappropriate since there is a long tradition of conceptual language that uses the expression *spirit* in some relation to intellectual activity. But the religious dimension of the concept of spirit is also important for the subject under discussion here. At the origins of human culture, as well as in individual development, the origins of language appear to be intimately connected to the origins of religious consciousness. If we are dealing here with the "field" within which human subjectivity, characterized by self-awareness, develops, such a field may be appropriately described by use of the term designating the concept of spirit.

The meaning of this word *spirit*, however, has become uniquely vague and opaque in our world of secular culture. The intellectual potential evoked by this word must, first of all, be explicitly reclaimed before we can use it again. Hence, it seems important, at least for a few moments, to take note of the intellectual spectrum that is associated with the word *spirit* and that resonates in its use.

In the history of Western philosophy, the concept of spirit has usually been more or less restricted to the phenomenon of consciousness. Thus, according to John Locke, we associate the word *spirit* with the ideas of thinking and willing "or a power of putting body into motion by thought and, which is consequent to it, liberty." By combining the notions of thinking and willing we achieve, according to Locke, "the idea of an immaterial spirit."[14] Therefore, Locke attributed those "activities of consciousness" to a "substance" he called "spirit."[15] This usage can be traced quite far back. As early as Augustine, *spiritus* was occasionally used as an equivalent for *mens*,[16] and, according to Thomas Aquinas, the human soul is called "spiritual" or "spirit" because of its intellectual potential.[17] But Thomas Aquinas also knew another and more complex meaning for the word *spirit*, a meaning that could also include material things and processes: namely, the word could express the intuition of an impulse or movement.[18] In a late echo of this broader understanding of the concept of spirit, in the eighteenth and early nineteenth centuries, the word *spirit* was occasionally used to describe the vivifying principle in all living things.[19] It was probably Hegel's thought, above all, that finally reduced the idea of the power of the spirit to the dynamic between concept and idea, and that was the starting point for the movement by which, after Hegel, the concept of spirit could be demythologized and reduced to individual consciousness.

The most important source in our culture for the broader notion of spirit, not reduced to consciousness, is surely the Bible, where spirit— the Spirit of God—is portrayed by analogy to the dynamic of the wind

(John 3:8; compare Gen. 1:2; Ezek. 37:9–10) and understood as the principle of life: according to Psalm 104, in the springtime, the Spirit of God renews the face of the earth, and all creatures must die when God takes from them the portion of spirit that is allotted to them (vv. 29–30). Similarly, the Yahwist narrative of the creation of the human being describes how God blew the "breath of life" into the nostrils of the clay figure he had formed (Gen. 2:7). In the view of the Yahwist writer, it was only this breath of life that made the human body a "living thing" or, in literal translation, a "living soul" (ibid.). Elsewhere, namely in Qohelet, this breath of life is quite explicitly identified with the divine Spirit (*ruah*): The spirit is given by God and is returned to God when the human being dies (Eccl. 12:7). Is it not quite natural here to recall the words of the psalm that, according to Luke, Jesus spoke as he died? "Father, into your hands I commend my spirit" (Luke 23:46 = Ps. 31:6). For our way of thinking, it is quite natural everywhere here to assume a clean separation between the Spirit of God and the human spirit, and that assumption has made its way, as a given, into Old Testament exegesis. But it seems that, as regards the spirit, the Old Testament texts do not suggest any such separation. Even the spirit in the creature, the spirit at work in the human being, remains ultimately God's property. The notion of the divine Spirit as the origin of all life found particularly powerful expression in the vision of the prophet Ezekiel of the resurrection of the dead nation, which so strikingly recalls the Genesis narrative of the origins of human life: The dry bones of the people of Israel clothe themselves with flesh and begin to live when the wind or spirit from God blows on them (Ezek. 37:5–6, 10; compare v. 14).[20] It seems appropriate also to read in this same vein the famous words of the apostle Paul in 1 Corinthians, according to which the first Adam (according to Gen. 2:7) was created only as a living being but the last Adam is himself a life-giving spirit: That is the reason the body of the risen LORD and those who will participate in his resurrection are called a "spiritual body" (1 Cor. 15:45). In the context of Old Testament views on spirit and life, the notion of a "spiritual body" can only mean that this will be a form of life that is no longer separated from the divine Spirit but rather lives in union with this, its origin. From this, it is a natural conclusion that

Paul expects the new life of the resurrection of the dead as immortal (1 Cor. 15:53ff.). Unlike the life hoped for in union with God and God's Spirit, this present life is not immortal because it does not remain in union with God, even though this life also owes its origin to the life-giving power of the divine Spirit. This seems to be the apostle's understanding of what it means to be a "living being" in the present order of things: It is true that such a being owes its existence to the divine Spirit, the giver of all life, but it is a reality separated from it, existing for itself, detached from this origin, and therefore mortal. If this interpretation of the Pauline statements is accurate, we should include within the field of its application also the distinctions Paul makes elsewhere between the divine Spirit and the human spirit, or between human reason and the Spirit of God (1 Thess. 5:23; 1 Cor. 2:11–12; compare 1 Cor. 14:14–15).

Paul's reinterpretation of the function of the divine Spirit in the context of biblical anthropology became the starting point and reference point for Patristic discussions of the theme. These took place in an intellectual climate shaped by Stoic and Platonic ideas. The Stoic notion of a divine *pneuma* that saturates the cosmos and finds its highest manifestation in the rational human mind overlapped with the biblical idea of the life-giving power of the divine Spirit; similarly, in the ideas of later Platonism about the divine character of the *Nus*, one could find a convergence with the Genesis account of the creation of the human soul by the breath of the divine Spirit. The Gnostics combined the Genesis story and Platonic notions of the soul by attributing to the human souls of the elect a participation in the divine nature of the *pneuma* from the moment of their creation. The church fathers, in contrast, regarded participation in the divine *pneuma* as an effect of the redemption and not of the original creation. According to them, the natural human being had no share in the divine Spirit, not even in the rational soul. Therefore, it is necessary for every human being to be reborn from the power of the Spirit if she or he is to attain to eternal life. Still, the fathers had to admit that the human mind was attuned to the divine Spirit even from its creation and that its life is grounded in the creative presence of the divine Spirit. But the human mind remains dependent on enlightenment by the divine Spirit, and its disposition for the Spirit is only fully realized through the redemption, the outpouring of the divine Spirit in the hearts of the faithful, whether this took place in the action of baptism, as Clement of Alexandria thought, or in the process of sanctification, which only began at baptism, according to Origen's teaching.[21] But even in Christians, to whom the divine Spirit is given as an enduring possession, that Spirit does not become part of their human nature but enlightens, motivates, and activates their lives through its divine power.[22]

Thus, in Christian Patristics, there always remains a difference between human consciousness, the human mind on the one hand, and the divine Spirit on the other. One must see in this an interpretation of the biblical words that, while it is affected by the confrontation with the Hellenistic divinization of the human soul and with Gnosis, is nevertheless restrictive, for those words speak less guardedly of the human soul's sharing in the power of the divine Spirit. Certainly that sharing is limited, both substantially and temporally, because human life, like the lives of other creatures, is subject to the power of death. Paul emphasizes that limitation, in contrast to the future life that will be enduringly united with the life-giving Spirit. Patristic theology emphasized these limitations on the natural human being's participation in the divine Spirit even more strongly because the church fathers rightly regarded the bestowal of the divine Spirit and its presence in the human being as a specific effect of the redemption. But it is questionable whether this did justice to the biblical conviction that created life itself has its source in the divine Spirit. The church fathers took this into account and spoke of a relatedness between the human soul and the divine Spirit, and they went beyond the Pauline statements by attributing this relatedness specifically to the human *nus*, the rational consciousness. Human consciousness, the human mind, thus not only has its origins in the working of the divine Spirit, as do all living things, but is disposed to receiving the enlightenment of the divine Spirit in various forms and stages up to the enduring indwelling of the Spirit in the human soul. This translated the biblical notions of the dependence of human beings, like all created life, on the divine Spirit into Hellenistic forms of thought; on the one hand, it retained a greater distance between created life and God, but on the other hand, it also attributed to that created life and the Spirit a greater degree of independence.

All these discussions are long since history. Can we still attribute to them any kind of relevance for the tasks of interpreting current experience? Can the dynamic association of human consciousness and divine Spirit still serve as a model, or at least a source of inspiration when we address today's problems of understanding and the description of the functions of consciousness, its origins, its relationships to the human body, and especially its basis in brain activity? That is just how theology should make use of biblical ideas and the Christian doctrinal constructions from the past: They are not to be treated as dogmatic definitions that prevent new developments in processing experience through thought but as a source of inspiration for an appropriate interpretation of contemporary experience. Certain ideas derived from this tradition, such as the notion of different parts of the soul, may be discredited by experiences that are now available to us. Nevertheless, theology will continue to hold to the Bible and the teachings from church tradition, trusting that even today, out of the riches of that tradition, new aids to orientation will continue to grow as we search for more appropriate solutions to the problems of current experience.

Before we took up the history of the concept of spirit in order to make certain of its tested latent potential for understanding the human soul, the human consciousness, it was already apparent what possible advantages the idea of the spirit as origin of consciousness could offer for the interpretation of the dependence of consciousness on cultural life in general and language in particular: Such an interpretation could help us to avoid the circle that emerges when we interpret the origins of consciousness from its dependence on culture and language, if culture and language, in turn, are understood, as usually happens, as products of consciousness. With the aid of the concept of spirit, we can name in culture and language that which is already present to consciousness and is not first of all a product of it. But precisely through the clarifications just derived from the history of the concept, it could appear doubtful whether the concept of spirit in its biblical origins has anything at all to do with the problems of consciousness, language, and culture. At any rate, the biblical notion of spirit applies primarily to the Bible's understanding of life and the question of its origins, not in the first place to human culture. Certainly, particular cultural achievements, especially the activities of artists, require, in the biblical view, a special degree of giftedness from the divine Spirit (Exod. 28:3; 31:3; 35:31). Such a special gifting from the divine Spirit is also presumed for the extraordinary achievements of the heroes and to make the kings able to carry out their duties. The special charism of the prophets—but also that of the poets and sages—is also to be reckoned within this category. The visions of the prophets, like their words, have their origin in divine inspiration. One must see this view of things not as something that simply falls outside our scope but in connection with those other special activities of the Spirit. What we seem to be dealing with are expressions of exalted, intensive vitality that, therefore, demand a special degree of communication of the divine Spirit of life. But do such extraordinary phenomena justify the generalized assertion that all cultural phenomena, and especially language, are traceable to the presence and activity of the divine Spirit in human beings? It may appear that the statement of the Yahwist story of Creation about Adam's being equipped with the life force from the divine Spirit represents such a generalization. Certainly, the narrative does speak of the breath of life. But the story of the creation of Adam does not apply the divine infusion of the breath of life directly to the phenomenon of language. Language is only mentioned later, when the narrative reports that God brings the animals God had created to the human being, as it says, "to see what he would call them; and whatever the man called every living creature, that was its name" (Gen. 2:19). Language is here described as a human invention, as Johann Gottfried Herder emphasized in his book on the origins of language, and not as a supernatural gift of God. On the other hand, however, one should not forget that history begins with the creation of the human soul by the divine breath of life. All the human soul's expressions of life are thus to be regarded as dependent on the working of the divine breath of life. Hence, human invention and divine inspiration cannot be regarded as mutually exclusive; rather, divine inspiration grounds and activates the spiritual abilities of the human soul.

In the present discussion, the relationship between language and religion seldom receives the attention this phenomenon truly deserves.

The dialogue between John Eccles and Karl Popper on consciousness and its origins is no exception in this regard. In connection with the cultural activities of human consciousness, they mention its myth-building activities once or twice, but they attribute no particular significance to them, as Ernst Cassirer had done in his theory of symbolic forms. Cassirer regarded magic and myth as the origins of language, although language, as we know it, no longer has the character of a magical invocation of reality. The mythic origins of language seem to have been of special importance for the descriptive functions of human speech, which Popper emphasizes as the specifically human element in it. Why does the descriptive function of human language have to do with myth? The naming of an object is originally an ecstatic experience because the object itself is experienced as present in its name.[23] Cassirer's thesis gained plausibility through the empirical results of certain of Jean Piaget's researches on the process of acquiring language in the intellectual development of children. Piaget discovered that the early development of linguistic ability was intimately related to the development of children's play, especially symbolic play, where the real object of the child's play is only represented by the toy, so that one can say that the child at play is ecstatically present to the object, the thing that is invoked through the noises made during play. Piaget identified "mythic" and what he called "animistic" elements in the children's intuitive thought and spontaneous utterance up to the seventh year of life.[24] In addition, he emphasized the function of these phenomena in the development of consciousness of an objective and self-sustaining world, within which then the child's own body and "I" are localized. According to Piaget, the first form of such objectivity is that of mythos in the emotional experience of play. All these observations are the more remarkable since, in this connection, Piaget neither refers to Cassirer nor reveals any sort of interest in the development of anything like a theory of religious origins of language. His theoretical construction was aimed in a quite different direction because he assumes as a matter of course the current attribution of behavior to a subject and, therefore, describes children's behavior as though the child were an independent subject almost from the day of its birth. Only under these assumptions could all experiences be explained as the activities of such a subject. But if, as Karl Popper assumes, the human soul is not a separate substance, not a subject of its experiences from the moment of birth, but only comes to be in the process of discovering the surrounding world and especially the social and cultural world, then the origins of human subjectivity must themselves lie within this process. Human consciousness is then not to be attributed to an already existent underlying subject but rather is to be understood as a process comparable to the phenomenon of fire, which nourishes itself from the combustible material around it. The consequence for our understanding of language is that mythic and religious spirituality in the process of the acquisition of language acquires a fundamental significance for the question of the constitution of human subjectivity. Such a religious spirituality may well have furnished the context for the first beginnings of language in human history.[25] Certain unique features of language, especially in its descriptive function, are better explained from such a perspective than on the basis of the assumption that language was developed primarily in the service of toolmaking or hunting.[26]

The biblical concept of the divine Spirit as the origin of the soul and thus also of consciousness is of special interest in reference to the problem of recognizing the truth of things, which is closely related to the descriptive function of language: How can we understand the fact that human consciousness and language are capable of grasping the nature of things? The possibility that human propositions can be true would present great difficulties if human consciousness were entirely passive and pure receptivity in its perceptions. But today we know that it is quite otherwise: Consciousness and the brain are most active at the moment when something is experienced, beginning with sense perception. How, then, is it possible that information we receive is not hopelessly falsified? On the basis of the biblical view of spirit and consciousness, we could answer that the possibility of grasping reality external to ourselves with our consciousness is founded in the fact that the Spirit in which we participate is also the origin of all life external to us, the origin of all the different forms of created reality. Something of the sort may underlie the curious statement in the Yahwist story of Creation that the "name" of every living creature was to be whatever the human being called it (Gen. 2:19). If we recall that, for the archaic mind, the name of a thing is not something external to it but contains the nature of the thing itself, it becomes clear that this biblical passage says nothing less than that the human being, because of its participation in the divine Spirit, is capable of grasping the nature of things. It is important here that the Spirit is not only the origin of consciousness but the origin of all life. Only because the Spirit grounds all created reality can it also be the origin of a consciousness that grasps things as what they are in themselves.

The ability of consciousness to perceive things is certainly not limited to living creatures but extends to all reality, although it may have a special connection to the nature of organisms because the human being is itself a "living soul" (Gen. 2:7). The reasons for the comprehensibility of inorganic reality may be analogous: Inorganic reality also has its origin in the divine creative Spirit, although it is not internally ensouled by that Spirit as are the living creatures. Such suggestions accommodate the findings of today's natural scientists, according to which modern physics no longer offers a materialistic description of the universe. In reference to the development of physics from classical mechanics, with its fundamental concept of the body, to the rise of the idea of field as the fundamental concept of modern physics and its attempts to derive matter itself from nonmaterial states, Karl Popper also says, "Materialism transcends itself."[27] On the basis of similar considerations, theoretical physicist Georg Süßmann writes, "Thus the material of all things appears to be as if crafted out of thought."[28] This should not be understood as expressing a philosophical idealism. Instead, it is that accepting a spiritual dynamic as the basis for the processes of nature makes it possible to understand the fact that human consciousness is capable of grasping the structure of these processes and making itself their master.

Süßmann sees no difficulties in accepting a correlation between spirit and life, on the condition that one begins with a sufficiently broad concept of life. He distinguishes three stages or steps of life, which, in the nature of things, represent three stages of internalization of the spiritual dynamic at work in things. He does not begin with plants and animals but with phenomena of currents and flames. He locates the vegetative and sensate life of plants and animals only at a second stage and then places at a third stage the intelligent life of human consciousness.[29] But what does every living being have in common with human consciousness, to make it more comprehensible that consciousness is capable of grasping the nature of things?

It may be that such a characteristic can be found in the feature of wholeness or form (*Gestalt*). Every living thing is itself a form, and it grasps other things, to the extent possible, from the point of view of their forms or shapes. This begins with very primitive types of perception of forms, investigating the environment of the living thing simply for the appearance of some less abstract features. Such behavior appears to be grounded in species-specific schemata of perception, and the appearance of features corresponding to these schemata evokes similarly species-specific responses. But at later stages of development, the qualities of forms are also distinguished by means of learning processes, and the act of perception can then also be unspecifically prepared for the elements essential to a particular form. Most of the kinds of forms we perceive are grasped in abstraction from time, but there is also a perception of living things whose form essentially includes the manner in which they move and develop within time. Every idea of an animal or plant involves such a living form, which is inseparable from the manner of its becoming and passing away. It seems to be a special characteristic of living forms that time and movement are not something accidentally added to them. Every animal is such a living form, although not all animals appear to be capable of grasping living forms in the way that is characteristic of human consciousness. The human being, however, not only understands animals as living forms but also plants and even inorganic phenomena such as a flame. All apparently active and self-determined systems we perceive, instinctively, as living, and it is well known that children experience their environment in this way to a much higher degree than do adults.

Let us add that all living forms are open systems. They are open systems because the continuation of their life processes, just as in the case of fire, is dependent on their environment, and animals, at least, have a relationship to the environment on which they depend that is grounded in the structure of their life-form, and so also to time, to the future of their own life, that will mean its transformation, even though the animal may not be aware, as such, of the future toward which its instincts are driving it. It is part of the living form of an animal that, in living its life, it internalizes time, incorporating its future transformation into its own life-form, although that future at the same time transcends its present state. Life is therefore characterized essentially by surpassing the self, by self-transcendence, and as soon as a living thing is capable of becoming aware of a living form in its temporal structure and orientation, and also perceives itself as a living form, as is true of human consciousness, it is necessarily also aware of the transcendence of time beyond its own life and death.

It is interesting that the Hebrew language has developed a word that very precisely expresses this self-transcendence and simultaneous neediness of the living thing. The word is *nephesh*.[30] The connotations of this word are largely lost when it is translated "soul." It is, of all things, the *nephesh*, the living thing in the process of its self-transcendence, that is described in the Yahwist creation narrative as the specific creation of the spirit. Accordingly, the spirit is said to have something to do with the production of form and wholeness, more precisely with the open system of the living form in the self-transcendence of its life-process.

Where have these reflections brought us? Do they yield any answers to our question about what the structure of a living being has in common with the activity of human consciousness, so that we might get a glimpse of their common rootedness in the dynamic of spirit? First of all, we can say of human perceptive consciousness that it grasps not sense data but impressions of forms, of distinct wholes. Beyond that, the analytic capabilities of consciousness make it possible to grasp forms as wholes that integrate their own parts and thus also living forms that continually integrate the conditions of their ongoing lives. Consciousness itself has often been described as an integrative, synthetic activity that is, nevertheless, connected to the ability to differentiate and thus to analyze.[31] Kant emphasized the synthetic and dynamic nature of perception. As an open system, it corresponds to the structure of the life process itself. But at the level of human consciousness, the process of continuing and self-surpassing integration is accomplished only in a field of awareness, an awareness that in more primitive forms of perception appears to have a more momentary character and to be limited to recognizing or re-recognizing abstract schemata in the environment, whereas at the level of human experiences, the forms grasped are perceived each time as wholes made up of parts and therefore also in turn as parts of larger wholes, as elements in a situation and, beyond that, of a cosmos extended in space and time. The human being learns to know its own body and the names applied to it as related to a place in the cosmos of the social and natural world and learns also to regard its own life as a process limited in time; inevitably this connects to the question of a future beyond these limitations, beyond the boundary of death, and of the powers that extend beyond and set in motion the objects of this world of mere appearances. This process of integration that goes beyond the given seems to unite human consciousness in the most intimate fashion with the dynamic of all life and reveals its spiritual nature, the participation of consciousness in its spiritual dynamic, which makes its life-processes possible by transcending them.

One aspect of such transcendence of the spiritual dynamic even beyond individual life and consciousness is expressed in familiar phrases about the "spirit" or "soul" of a community. Human community in its various forms offers the most likely, though not the only, example of processes of spiritual integration extending beyond the life of the individual. But on the other hand, community life is never characterized only by the aspect of integration. Although, within a community, every individual member participates in its spiritual dynamic, there are tensions and antagonisms in community life between the individual members. How can this circumstance be applied to a pneumatological interpretation of human reality?

In biblical tradition, as in other archaic cultures, there is a notion of evil spirits. How is it possible for a spiritual phenomenon to be evil? This can be described in the language of the integrative dynamic of spiritual processes thus: Participation in the spiritual dynamic of life is always predicated of a living form that has internalized that dynamic, which is organized around a center of its self-transcending activity. But when the self-centeredness of a living process dominates over the dynamic of self-transcendence, so that the living being can no longer be a member of a larger spiritual integration, the dynamic of self-transcending integration itself becomes a principle of separation and opposition. In this lies not only the root of the possibility of evil but also a perspective from which we can achieve a better understanding of the early Christian distinction between the divine Spirit and the human spirit, or human consciousness: Although every living thing has a share of the life-giving breath of the divine Spirit and lives only through it, no living thing as such is identical with the dynamic of the Spirit because every living thing, in its self-centeredness, can become evil. This is true even of human consciousness, although—or perhaps even because— the spiritual dynamic is internalized to the highest degree in human consciousness. Therefore the self-centering of human egoism can turn against the life-giving working of the Spirit in an especially destructive way. But for that very reason, on the other hand, the human being is shaped by a desire for fuller participation in the Spirit, which would satisfy its hunger for wholeness and identity and bring it peace with all creation. But the fulfillment of such a longing is not given to the human being in the form of a definitive possession: Such fulfillment would enable human consciousness inevitably to exceed its own self-transcendence toward another. Therefore the longing finds fulfillment only in the ecstatic experience of faith and its hope, and in the creative love that is born of such faith.

# The Human Being as Person

According to Immanuel Kant, the three chief questions of philosophy converge in the question of humanity: What can I know?
What should I do? What may I hope? But the subject of anthropology
also links together the great circle of the human sciences, especially the
three higher faculties of the medieval university: jurisprudence, medicine, and theology. For the efforts of the physician, the lawyer, and the
judge, as well as those of the pastor, are all on behalf of the human
being. All of them—as well as the teacher and social worker—need a
sufficiently broad and complex conception of the human so that, in
their professional practice, they will not see human beings in a narrow
perspective and treat them, accordingly, in a one-sided manner. Hence,
the question of the human is especially appropriate for interdisciplin-
ary dialogue.

Likewise, the human personality deserves a special status among
the many and varied themes of anthropology. Is it not true that, for
us, the concept of person is bound up not only with the concrete reality of the human being as individual but also with the special dignity
and inviolability we attribute to the human individual? This point of
view directs us first of all to the sphere of law and rights. As persons,
human beings possess fundamental rights or human rights. That is why,
in the constitution of the Federal Republic of Germany, the rights of
the person head the catalog of fundamental rights (Article 2): the right

to free development of the "personality," the right to bodily integrity, and the right to freedom of the "person." This personal freedom is said to be "inviolable." It is thus intimately connected to the "dignity of the human being," which Article 1 describes as "unassailable" and which, according to Article 1, Paragraph 2, constitutes the basis for human rights.

Thus, the human personality is closely connected to the unique freedom and dignity we attribute to every human being. But what constitutes human personhood, to which such freedom and dignity are assigned? On this point, views diverge widely. In 1966, the Berlin philosopher Michael Theunissen gave an overview of the different ideas of human personality in twentieth-century thought; in doing so, he emphasized particularly the contrast between the authors who regard the social relationships in which the individual lives as constitutive of personhood and others who see personality as anchored particularly in the self-identity of the individual, prior to all social relationships, and who even interpret it as the human "through-and-by-oneselfness."[1] Still, these divergent notions have a common basis in the fact that the concept of person is always applied to the concrete individual. The same is true for anthropological medicine, whose concept of person (Paul Christian is an example)[2] emphasizes above all the body-soul unity of the individual.

But to what extent is the individual human being really a person, the subject of that unique freedom and dignity, with his or her associated aura of inviolability, of which our constitution speaks? To put it another way: From what point on is the human being a person in this sense, the subject of human rights? Is it from the moment of birth or even before birth? As we all know, this is a highly disputed topic today in the context of the arguments about the juridical permissibility of abortion. Hence, the answer to the question of when a human being becomes a person in the sense intended by our legal system has far-reaching consequences. All the more understandable, then, is the urgency with which people have sought applicable human biological or medical criteria. Where is the threshold beyond which the human fetus is a person and, as such, inviolable? Is there such a threshold, or is there not, rather, a continual process by which the child develops in its mother's womb and that provides no convenient empirical criteria by which we can decide the question of the point at which the enormous step is taken that makes the fetus into a person? But if there is no such empirically identifiable developmental leap of that kind, can our concept of person itself be in any way substantiated?

Similar questions are associated with individual life in the phases of its decline, dissolution, and end. Is the human being who will never again awaken from a coma still a person? Anyone who follows the idealist tradition and sees consciousness and self-awareness as constitutive for the concept of person must be rendered uncertain at this point. And yet there are good reasons for hesitating to give a negative answer to the question, not least because further questions immediately follow: Is someone with advanced memory loss, one who has lost long-term memory and the ability to speak, who can no longer recognize his or her closest relatives and appears incapable of personal contact with others still a person? Kant said, "The fact that man is aware of an ego-concept raises him infinitely above all other creatures living on earth. Because of this he is a person; and by virtue of this oneness of consciousness, he remains one and the same person despite all the vicissitudes which may befall him."[3] But a person affected by such advanced loss of memory that he or she has not only lost the ability to speak but can no longer recognize his or her closest relatives is far beyond possessing a "oneness of consciousness . . . despite all the vicissitudes which may befall him," because that is dependent on memory. It is, at least, very doubtful that he or she still "is aware of an ego-concept" as Kant says, for we now know that the origin and use of the word "I" is associated, within the life history of an individual, with the development of social relationships, and those are scarcely existent in the life of such sick persons. In addition, according to Kant, the notion of the "I" is only possible in connection with a unity of consciousness, something that ceases to be when there is advanced memory loss. It is, therefore, at least very doubtful whether there is still an "I" in such sick people. What we know about the conditions for an awareness of the self tends to speak against it. What, then, of the personhood of someone in that condition? If the personhood that, according to Kant, "makes us a being infinitely elevated above" the nonrational animals is grounded in the fact of an "I" and a resulting unity of consciousness, it would seem that we must deny personhood to such sick people. The legal consequences would be very broad: Such a sufferer would then no longer enjoy the protection of the inviolability of his or her personal freedom. The right to life and individual integrity is also, in our constitution, understood as belonging to persons: that is, it depends on the concept of the person and, as to its possible application, the existence of a person.

There would seem to be problems similar to those concerning extremely advanced and irreversible loss of memory with regard also to mental illnesses that involve a disintegration of the unity of the self and of consciousness. I assume that most physicians, out of their own feelings and also for moral reasons, would be inclined to treat even such a sufferer as a person and thus as a human being. But what is the support for such an attitude? What is the foundation for the assumption that such a sick man or woman is still a person? The idea of the person as defined by the "I" is here no longer adequate. But that means that the foundations for our understanding of human personality must be more deeply laid; if, as the constitution says, human dignity is inviolable and if that inviolability is closely bound up with human personhood, then the concept of person must be such that it can be applied to everything that has a human face, including sick people whose ego-structure is destroyed, as well as to the newborn or still unborn child whose ego-structure is not yet developed.

What, then, makes the human individual a person? Let us return once again to the statements in the constitution about the inviolability and sanctity, freedom and human dignity of the person. What kind of concepts are these? Are they so characteristic of human beings that they can be found always and everywhere, in every human being? Hardly. It is true that we probably all know people who exude an unmistakable dignity, not something attached to an office but a dignity that is simply part of their humanity. And there are also examples of people who live, in the most impressive way, independently and sovereign over the external conditions of their existence, at least in certain situations. But can it be seriously maintained that such dignity and inner freedom belong to every human being in every life-situation? Physicians probably know more than other people how miserable human life often appears to be, not only in its physical but also in its spiritual reality. Moral wretchedness and misery are encountered more frequently than not in our daily lives. In all these cases, we are not talking only about borderline cases of physical helplessness, spiritual collapse, and criminal activity. None of us is perpetually in our highest physical, spiritual, and moral condition. The dignity and inner freedom attributed to the human being as such are very seldom to be inferred from the empirical reality of people in general. And yet the good physician will not treat the patient lying helpless before her or him simply as one being among others in need of help but will respect him or her in light of what the human being should and can be and also what this particular human being once was or may yet be. In that light, he or she is able to perceive in the helplessness of the patient that personal dignity that the condition of the patient in many cases does *not* reveal. The same is true, or should be true, for the attitude of the teacher to the student, the judge to the accused, the jailer to the convict. The same should also be true of the behavior of competitors toward one another, of opponents, victors and vanquished, superiors and subordinates, and in general of husband and wife, parents and children.

Thus, that the human being is a person is not merely a characteristic that can be demonstrated of every human individual, always and everywhere. This is especially clear in regard to the inviolability and sanctity attributed to the human being as person. Unfortunately, from a purely empirical point of view, neither life nor individual freedom is truly inviolable. People can be robbed not only of their lives but, through manipulation or illness, of their freedom as well. The statement that the human being is sacrosanct because of her or his dignity as a person is essentially counterfactual. It does not say what is obvious and always the case but should be. It is true that *ought* and *is* cannot be cleanly separated here. Human existence itself is shaped by this *ought*. Therefore, we can, in fact, simply formulate in the indicative: "The dignity of the human being is sacrosanct."

Such sanctity was not always part of the definition of a person. From its Greek and Latin roots, the word *prosopon* or *persona* meant, first of all, the face and beyond that, especially the mask worn by an actor. This theatrical meaning of the word *person* then led to its being used to designate the social "roles" people play, especially the socially prominent roles through which one becomes a public person. There is a very similar usage in the Bible: It is said that a judge may not "be a respecter of persons" (Deut. 1:17; compare 1 Pet. 1:17). Accordingly, it is said of God's incorruptible judgment that God "shows no partiality to nobles, nor regards the rich more than the poor" (Job 34:19). Here, as also in Greek and Roman usage, the word *person* thus also describes especially a high social standing. On the other hand, in the legal language of late antiquity, this word was used quite generally for individuals, for example, for various numbers of individuals. The word *person* could also stand for a single individual. But then it had acquired a generalized sense: The single person was here viewed only as an example of the category "individual" and precisely *not* as the subject of special dignity, certainly not of sanctity.

The idea of the inviolability of the human being as such, so closely connected to our current concept of the person, originates in the Old Testament, where it appears in connection with the creation of the human being in God's image. Because of being created in the image of God, the human being, every human being, shares in God's own sanctity. So, in the book of Genesis, the command not to murder is grounded in the fact that God created the human being in God's own image (Gen. 9:6). Thus, the origin of our idea of the inviolability of the human being as person lies in the biblical belief in Creation. The dignity belonging to every human being as such is founded in the divine intention for humanity. In other words, the statement about the human being's creation in the image of God must be read as a statement about human *destiny*. In the New Testament writings, Jesus is called the image of God. Of him it is said not only that he was created in the image of God, but Paul calls Jesus himself the image of the eternal and invisible God (2 Cor. 4:4; compare Col. 1:15). Moreover, all people are to bear the image, visible in him, of the human being coming from heaven, from God (1 Cor. 15:49); all are to be formed according to his image (Rom. 8:29). This means nothing less than that, in Jesus Christ, the destiny of humanity itself has been realized, the human destiny to make God visible in this world, to be God's image. The inviolable dignity of every human being is grounded in that destiny. In the history of Christology, then, this idea was combined with that of the person. The Council of Chalcedon, in 451 CE, wrote that, despite the difference between divine and human in Jesus Christ, his person is that of the Son of God: In the visage of Jesus Christ, we encounter God's very self, in that we recognize the Father in the obedience of the Son to his mission. This statement has not only christological but also general anthropological relevance if, in turn, we are all meant to bear the image of the human being that has appeared in Jesus Christ. In the person of Jesus Christ, Christians recognize the destiny of the human being to community with God, the basis for the personal dignity of every individual human being.

Absent this point of view, of such a human destiny infinitely above the frailty and wretchedness of our lives, there is scarcely any ground for high ideals about the sacred dignity of the human being and the freedom to which every human being, as person, is called. But if, in the knowledge of the divine destiny of the human being, we turn to individual people, we will discover even in their faces the traits of personal dignity and freedom that are founded on that destiny. Thus, the eye of love beholds in the beloved face more than the indifferent gaze of a passing stranger. But true love sees in the other not only the counterpart to one's own wishes; it also sees in him or her the features of that person's own special and distinctive calling to be what she or he should be as a human being, and therefore always already is in some way, because that *should* is the Creator's word about the human being.

It is true of each of us that we do not know what we will become, and yet we already are it in some way. I am still on the way to my own special self, and yet I am already and always have been myself. In the ecstasy of emotional life, we achieve a relationship to ourselves as correlated to the world that we learn to distinguish from ourselves. In doing so, we understand ourselves not only as others see us, in this particular physical existence, the object of social evaluations and expectations. The ecstasy of feeling goes beyond that. In it, world and self are still unseparated. The indefinite whole of life is present in emotion, and world and self emerge from it through the distinguishing activity of the rational mind. But something of that nameless whole remains connected for us, both to our view of the world and to the idea of our self. Thus, our relationship to the whole of our self is grounded in the life of the emotions, beyond everything we are in the eyes of others. In this relationship to our self, we have within our own selves a relationship to the more or less vaguely grasped destiny of our self, through which we are persons.

The construction of our identity is the concretizing of this relationship to the self, first and primarily given to us through feeling. The process of constructing our identity is about a more precise understanding of our self, an understanding that includes everything that belongs to our individual life. We should not understand the individual steps in the process of identity building as the achievements of the ego, as if our "I" were already definitively present and could decide about the content of its selfness. It is rather the reverse: The "I" taken for itself would be a thing only of a moment. The first-person singular, as linguistic analysis has shown, points only to the particular speaker at the particular moment. The identity of an "I" extending over time, by means of which it remains the same through the shifting moments of life, is not yet a given. The "I" owes its stability first of all to the identity of the self, which is achieved in the process of identity building. This is a better way of understanding the relationship between our personhood and our ego. We experience ourselves as persons in that our destiny as human beings, our selfness, toward the fullness of which we are always moving, is yet always present and appears in our "I." What we really are, from God, appears to us, broken in the mirror of our earthly life history, in the moment of the "I," and even and also there where the "I" knows itself to be painfully separated from its own self or where it suppresses the voice of its self and desperately, against that voice, wills to be or wills not to be itself, as Søren Kierkegaard so penetratingly described.[4] All that, even in its perversion, is only the forms of the presence of the destiny beyond our "I," to be ourselves. That destiny is present to us in the moment of our "I." That the human being is a person is thus, in fact, not founded in the ego. Our ego-consciousness is only the place where our selfness appears to us, as it appears to us in our faces. It is not without some deeper sense that it is precisely the *face* that is the starting point for the history of the concept of the word *person*.

# The Human Being as Person

In the introductory remarks to this brief essay, I attempted to show that the traditional understanding of the person in terms of the "I" as the subject of the unified consciousness is not adequate to explain our ideas of the dignity and sanctity that, for us, are associated with the idea of a person. In particular, such a conception does not permit us to consider as persons even people who have either not yet developed a stable ego structure or in whom there seems no longer to be one. But if we understand the person in light of the divine destiny of each individual to his or her true selfhood, which at every moment infinitely surpasses human existence and only becomes visible to us in the presence of the "I," then there are good reasons to consider as personal even those stages of human development in which no "I" has yet been structured and to count as persons as well those people whose "I" identity has been more or less destroyed. In them, too, reverence for the destiny of the human being will still recognize the dignity of the human person.

# Aggression and the Theological
# Doctrine of Sin

IN 1963, in the title of one of his most successful books, *On Aggression,* Konrad Lorenz described aggression as "the so-called evil," indicating that aggression is something that is generally regarded as evil.[1] At the same time, however, the German title of the book indicates that this judgment needs to be revised: Aggression is not evil as such but only "so-called" evil. Indeed, the book is meant to show that the instinctive roots of aggression were originally important for survival and are limited in their effects. According to Lorenz, aggression is "an instinct like any other and in natural conditions it helps just as much as any other to ensure the survival of the individual and the species." It is only our distance from the natural conditions of the life of our species that have released the "evil effects" of aggression among humans.

If we ask who it was that designated aggression as evil, so that he can speak of a "so-called" evil, Lorenz pointed to Sigmund Freud's theory of the death wish.[2] In fact, Freud, who stands at the beginning of the twentieth-century discussion of the psychological phenomenon of aggression, in his book on the discontents of culture from 1930 identified aggression, this "derivative and principal instance of the death wish (110)" with evil. The discovery of a human inclination to aggression caused Freud to judge the assumption that human beings are by nature "simply good" to be an "unsupported illusion (103)." That was also Freud's main objection to communism because according to that theory, it was only the "institution of private property" that corrupted human nature. Freud also laid the responsibility for the rejection he uncovered in positing a death wish at the portal of the broad inclination to suppress from consciousness the ugly side of human nature that reveals itself in aggression: "For the infants do not like to hear mention of the inborn human inclination to 'evil,' to aggression, destruction, and therefore also to cruelty (108)." Freud's proximity to the theological doctrine of sin, emphasized by Paul Tillich, receives especially clear expression in this statement.[3] The same is true of the contrast between aggression, as an inclination to damage one's fellow human beings, and the love commandment, a contrast that, in Freud's description, characterizes the evil quality of aggression, which "is made necessary at culture's expense," for "as a result of this primary enmity among human beings, societal culture is constantly threatened with destruction (102)." And according to Freud it is "the fateful question for the human species" whether it will succeed "in mastering the destruction of common life by human drives to aggression and self-destruction" with the aid of the "eternal eros (129)."

Freud was not the first to describe wickedness as an inclination to harm others: Arthur Schopenhauer had earlier said of it that—unlike mere egoism—it "seeks, quite disinterestedly, the hurt and suffering of others, without any advantage to itself."[4] But on the other hand, pure destructiveness, utterly purposeless malice, is seen as impossible for human beings. Immanuel Kant had already emphasized this view, but it also corresponds to the view of Christian tradition. Only in the idea of Satan is such extreme malice proposed as something real. Even in the case of Satan, however, Christian theology does not see destructiveness as the original root of evil.[5] This is probably related to the fact that Christian theology did not conceive Satan in the sense of a mythological dualism as the original counterforce to divine love but instead as a fallen creature of God. The root of Satan's malice, according to Augustine, was his inability to make himself God. In Augustine's view, the unlimited self-love that seeks to set oneself in place of God results in a hatred of God that turns against all that God has created. Destructiveness is the consequence and visible form, not the origin of Satan's malice.

In the perspective of Christian teaching, still less than in the case of Satan, can there be an original inclination to pure destructiveness in a human being. Human sin is not, as with Satan, naked self-love, a limitless *amor sui*, the immediate consequence of which is hatred of God, *odium Dei*. Human beings, as sensate creatures, are oriented to the objects in their own world. Their sin is greed, the desire to possess and enjoy things, and only implicitly—that is, in a veiled form—is unbridled self-love present in human lust (*concupiscentia*) as the ultimate driving force: Lust seeks things not for themselves, as goals in themselves, but as means for the pleasure of the one who desires them. Therefore, lust always presumes self-love. A person driven by lusts has always already placed himself or herself in the place that belongs to God alone because God alone is the highest good for creatures, the goal to be sought for itself alone. To that extent, in a latent manner, hatred of God is present even in human sin in the excess of self-love but only latently. For that reason alone, also, human sin can be healed; at any rate, God can heal it by fulfilling the human longing for happiness with his, God's own presence, and so with God's own love for his creatures.

The classic position of the Augustinian analysis of sin—which does not directly imply the additional hypothesis of inheritance but is the basis of the doctrine of original sin itself—is grounded in the fact that Augustine's reflections only make explicit what is implied by the apostle Paul's summary description of sin. In the letter to the Romans, Paul summarizes the meaning of all the requirements of the Law in one prohibition: "You shall not covet!" (Rom. 7:7), and correspondingly he summarily describes sin as lust. Augustine only uncovered the logical implications of the phenomenon of lust in his doctrine that lust always presupposes a purpose or end different from the particular object of desire and that the one who desires always wishes to make himself or herself that ultimate end, with the consequence that unbridled egocentrism, *amor sui*, is the real root of sin.[6]

The Augustinian doctrine of sin has been accused of finding concupiscence primarily manifested in human sexuality, which formed the starting point for a fateful development, a long history of Christian suppression as well as negative loading of sexuality.[7] In fact, Augustine's thought on this question, as is true of other ancient Christian authors, was deeply influenced by ascetic tendencies in the spirituality of late antiquity. But the foundations of his concept of sin as concupiscence are independent of that. In essence, these statements by Augustine about the nature of sin are a formal analysis of a "reversal of the motivating forces" of action, to speak in terms of Kant's doctrine of radical evil, which is analogous to Augustine's on this point. The fundamental idea is that it belongs to God alone, who is both goal and origin of the world order, to be sought by creatures as the ultimate goal of their actions. Everything else must be valued and used as a means on the way to God as the highest good. This order of the universe is perverted when human beings use the things of the world not as means to God but as means to support their own self-love. Thus, people exchange means and end, *uti* and *frui,* by making themselves the end instead of the means in relationship to God and misuse God, no differently from the things of this world, as a mere means to their own pleasure. This analysis of sin as a reversal of the means-end order of the universe is free of any narrowing down to sexuality alone. In Augustine's eyes, however, sexuality appeared to be an especially obvious example of this generalized perversion of human behavior, and the ascetic tendencies in late antiquity led him astray, so that, in many of his writings, he identified the perversion of sexual behavior with that sphere of action altogether. Augustine's theory of the inheritance of sin contributed decisively to this tendency and to its fateful effects throughout the history of Western Christianity because Augustine regarded sexual desire and lust as the means by which sinful concupiscence was transferred to posterity. However, we should keep in mind that Augustine's theory of the inheritance of sin functioned only as a secondary hypothesis to his structural analysis of the nature of sin. He proposed it in order to preserve human responsibility for sin. Because, if one is responsible only for actions that could have been prevented, then with a knowledge of the way human behavior is structured by egocentricity and concupiscence the question must arise: Is the distortion of existence described by the concept of sin to be regarded as in any way a matter of human guilt, or is it not rather an unavoidable fate? Augustine's answer to that question was that, while the actions of people now living are always predetermined by concupiscence, Adam, the first human being and likewise the embodiment of the whole race, had disobeyed the divine command by his free decision. That is how the distorted relationship to the things of the world, perverse concupiscence, came about. But for present humanity, Adam is not merely some individual from an earlier generation; because, as the first human being, he at the same time embodied the nature of humanity as such, all later people share in his story. To make this participation more anthropologically concrete, Augustine proposed his theory of the inheritance of sin, which had already been prepared for by Tertullian. The transmission of Adam's sin to his posterity by means of the concupiscence that was active in their begetting served Augustine as a guarantee of their participation in Adam's fall and his responsibility. It is not necessary here to enter into all the arguments that have been developed in modern times against Augustine's special hypothesis about the inheritance of sin. It is enough to say that this hypothesis, which is so closely identified with the stigmatizing of sexuality, has been altogether abandoned in current theology—in Protestantism, at any rate.

It was necessary to examine the Augustinian link between sin and sexuality at such length because we should probably see there an introversion of aggression, a case of self-destructive shifting of aggressiveness inward. The Christian doctrine of sin not only reveals much about the roots of aggression; it is, in some of its most influential forms, Augustine's probably the most prominent among them, itself a manifestation of this phenomenon. Thus, the Christian doctrine of sin belongs within the group of examples that demonstrate that Christianity itself has produced aggression, even though its message about the redeeming love of God proclaims the overcoming of sin.

Only when the motif-layer of self-aggression has been cleared out of the traditional Christian doctrine of sin can we—as above in the sharp distinction drawn between Augustine's description of the nature of sin and his subordinate hypothesis about its hereditary character—reveal the layer in which we can understand how sin is the root of aggression.

Certainly, in the concept of original sin, the sinfulness that structures human behavior is often associated, without distinction, with the special notion of the inheritance of sin. That makes it understandable that the whole topic of sin and guilt could be regarded as an expression of self-aggression, as in Friedrich Nietzsche's critique of traditional Christian teaching about sin. For him "the bad conscience" was seen as "the serious illness that man was bound to contract under the stress of the most fundamental change he ever experienced—that change which occurred when he found himself finally enclosed within the walls of society and of peace." What, according to Nietzsche, was this sickness? "All instincts that do not discharge themselves outwardly *turn inward*. . . . Hostility, cruelty, joy in persecuting, attacking, in change, in destruction—all this turned against the possessors of such instincts: *that* is the origin of the 'bad conscience.'"[8] Decades later, then, Alfred Adler and Sigmund Freud described the genesis of conscience in self-aggression in similar fashion. In the 1887 *Genealogie der Moral*, Nietzsche had already anticipated the derivation of the idea of God from "*fear* of the ancestor and his power" and "consciousness of indebtedness to him,"[9] a thesis that was presented twenty-five years later by Freud in *Totem and Taboo* (1912). And as, on the one hand, he found the self-aggression of guilt feelings to be the origin of belief in God, on the other hand, he assumed an intensifying reverse effect of belief in God on guilt feelings: "The advent of the Christian God, as the maximum God attained so far, was therefore accompanied by the maximum feeling of guilty indebtedness on earth."[10] Thus for Nietzsche, not only the phenomenon of the judgment of conscience and the idea of guilt toward the deity, but, closely connected to it, the idea of God itself, culminating in Christian belief in God, were all interpreted as the products of internalized aggression. Hence for him, atheism became the promise of liberation from such an oppressive burden. The underlying presupposition for this "hypothesis" of Nietzsche about the origins of "bad conscience," the idea of God, and the notion of "guilt toward the deity" in a "sickness" of the original "animal soul" of the human being, of course, is that this original animal situation out of which the history of humanity had begun was a condition of health and integrity. Absent that presupposition, a "bad conscience" still remains an indicator of a nonidentity in the human being, but such a nonidentity is no longer to be interpreted as the loss of a previous identity—something that remained with Nietzsche as a remnant of the Christian doctrine of original blessedness in secularized form—but rather, the awareness of nonidentity must be understood as a sign that we know that the identity of the human being is something not yet realized. Thus, the "bad conscience" becomes the reverse side of the situation that humanity has, from an early stage, conceived its identity through a bold projection beyond its factual status. And the history of religious experience of divine power is then to be understood as the history of a path toward freedom rather than, with Nietzsche, as the history of a sickness, a loss of original self-identity. The experience of guilt and the concept of sin are then, in principle, to be understood as realistic expressions of the insight that the human being is not yet identical with the idea of his or her destiny. The awareness of such a nonidentity is then not the product of a self-aggression but the realistic verso of the awareness of one's destiny, gained in the human action of religious self-transcendence. Therefore, bound up with the experience of religious liberation and elevation of the human being beyond what he or she already is to a greater insight into his or her existence is necessarily an awareness of the distance by which one is still removed, by one's own action, from this, one's divine destiny. And at the same time, the awareness of such a destiny as one's own is the ground for accepting responsibility for the condition of one's own existence and behavior, still so far removed from this goal, that is, an awareness of *guilt* in view of what should be. It is right that human beings have, as a rule, perceived this human destiny as a social one, so that religious experience is simultaneously an experience of the unity of society. To that extent, Nietzsche was correct—and Freud went in the same direction—in supposing that conscience and human socialization are related. But society is not only an entity making demands on the individual from outside and driving him or her by its demands into the corner of self-aggression; society is also the locus of his or her possible identity. This is founded in the fact that, in religious teaching about human life, the individual and society are inextricably bound up with one another.

If the awareness of guilt and failure to achieve one's own human destiny is not to be seen as, in its roots, the product of aggression turned inward, it nevertheless can easily join with such self-aggression: It is but a single step from knowing one's own nonidentity to self-hatred. Pride does not easily bear with a consciousness of one's nonidentity and failure to be one's full self. But if it is abandoning the proud identification of one's own existence with the full accomplishment of one's calling that facilitates the step from the awareness of one's nonidentity to self-hatred, it is also clear that this step toward self-aggression is just as much an expression of sin as is aggression toward others. In both cases, the starting point is an awareness of the unattainability of equality with God, something of which our pride longs to boast.

And so our reflection returns us to the Christian doctrine of sin as an interpretation of the origins of aggression. That aggression not only against others but also against one's own self is an expression of sin was, however, not yet observed in the classical development of the Christian doctrine of sin by Augustine. Otherwise, Augustine would surely have been more alert to the influence of motifs of self-aggression in his own thought, especially in the stigmatizing of sexuality that marks his doctrine of sin. It is true that Augustine—and the whole tradition that followed him—explicitly condemned the most extreme form of aggression against the self—namely, suicide—as sin. But the reasons for this—that suicide is also murder, that, as in the case of Judas, it expresses despair of God's mercy and leaves the one committing it no further chance to do penance[11]—do not connect the psychology of suicide with psychological analysis of the nature of sin. From another point of view Augustine was certainly aware of the connection between the sin of pride and the will to destruction. But he understood it only in terms of the very ancient concept of envy, going all the way back to Plato: The prideful angel Lucifer was envious, as a consequence of his pride, which separated him from God, so that, in his tyrannical pride, he enjoys the subjection of others rather than subjecting himself to God. And in his envy, he could not bear it that human beings should remain innocent; that is why he seduced them.[12] Augustine's doctrine of sin could be adduced as proof of Helmut Schoeck's thesis that *envy* is the "basic phenomenon" that underlies such forms of behavior as "aggression, hostility, conflict, frustration," etc.[13] But Augustine's viewing all this from the standpoint of envy blocked him from perceiving the phenomenon of aggression turned inward. It is true that Augustine emphasizes that even God's enemies cannot do God any harm but, in their fall, only damage themselves.[14] But this does not lead him to the insight that the sinner's pride can result even in the *intention* to harm himself or herself, and not only to harm others. The phenomenon of self-aggression thus shows that the concept of envy is not sufficient to explain the relationship between sin and aggression in its full profundity. Hence, one should perhaps view envy rather as a form of expression of aggression along with others rather than the reverse, as the root of all aggression.

The later development of the theology of sin, as regards the relationship between sin and the deliberate intent to do harm to God's creation, has not moved beyond the Augustinian concept of envy. But there has been some important preliminary work toward a clarification of the problems Augustine left unresolved. First we should mention Søren Kierkegaard's description of the relationship between anxiety (*Angst*) and sin. It is true that, in his 1844 book, *The Concept of Anxiety*, Kierkegaard did not make an explicit connection between *anxiety* and *aggression*, but that connection is present in modern discussions of the aggressive complex.[15] The frustration that is at the center of one of the most important hypotheses for the psychological interpretation of aggression[16] not only creates anxiety but is itself, even before its appearance, the object of an existential *Angst* that precedes all frustration, containing an indeterminate knowledge of one's own vulnerability and peril.[17] To that extent, then, anxiety regarding one's own potential to be and to endure is bound up with the experience of frustration. Even though, nowadays, frustration can no longer be regarded as the sole cause of aggression—the psychology of learning has shown the significance of imitation of aggressive behaviors, especially the disappearance of sanctions or even the rewarding of aggressive behavior by the group—nevertheless, the significance of frustration as an essential factor in the creation of aggressive behavior is indisputable. Certainly not every frustration leads to aggression, either immediately or in the longer run through the buildup of aggression. Frustrations can also lead to depression, or they may be the starting point for a positive change in behavior, a constructive adaptation to social conditions. Thus, other factors must be added if frustration is to express itself in aggression, factors of individual disposition and social conditions surrounding and affecting individual behavior. Only one-sided theories of aggressive drives can overlook all this in favor of an oversimplified path from frustration to aggression. On the other hand, there may well be aggression without any relationship to definite, demonstrable frustrations, inspired rather by bad examples and rewards for imitating such behavior. But even in such cases, we must presume anxiety as the general starting point for aggression. Anxiety derived from frustration promotes in some cases a turn to flight or depression but, in others, a move toward aggression.[18] However, the existential *Angst* that precedes the experience of frustration must be assumed as the starting point even for aggressive actions that are not provoked by frustration. *Angst* regarding one's ability to assert one's own social value may well play an essential role especially in the imitation of aggressive examples.[19] Thus, in any case, *Angst* must be regarded as the general precondition for aggressive behavior, even though the shift from anxiety to aggression is dependent on additional elements. The association between anxiety and aggression, however, reinforces the relevance of the description of the connection between anxiety and sin, as developed by Kierkegaard, for the topic of aggression.

Kierkegaard's guiding interest in his approach to the doctrine of sin was, remarkably enough, his effort to defend the reliability of the biblical portrayal of the original perfection of the first human beings against modern critics. Friedrich Schleiermacher, in particular, in his dogmatic theology, had objected to the notion that the sinful condition of present humanity could have followed a preceding condition of innocence, saying it was impossible to understand psychologically how the sin of Adam and Eve could have occurred, as portrayed in the biblical story of the Fall, "without sinfulness being already present."[20] For if Eve lent her ear to the whispers of the serpent and if Adam ate of the apple given him, there must already have been an "inclination to sin" there. Against this, Kierkegaard thought he could nonetheless suggest a psychological motive that was itself not yet sinful, not yet directed against God and his commandment, and yet provided the psychological "intermediate condition" for the transition from innocence to sin.[21] He thought he had found this psychological intermediate condition in anxiety, for, according to Kierkegaard, *Angst* must be distinguished from fear that is directed at particular objects. The indeterminacy of *Angst* shows that human beings fear primarily for themselves, namely, for their personal unity.[22] To fully understand this idea—that is, in light of the question of how and why human beings must be *anxious* about their personal unity, their personal identity—we must also look at Kierkegaard's words in a book written five years later, *The Sickness unto Death*. According to this work, human beings are constituted by their relationship to the infinite, and they know themselves to be so related to the infinite. But although they are aware of themselves, they cannot be self-determining, self-realizing, because their existence, as a relationship to the infinite, can only be realized by the infinite God. Therefore, human beings lose themselves when they try to be self-grounding (as distinguished from being grounded in God). But since, in their self-awareness, human beings always have a relationship to themselves, they are anxious about that self. This anxiety is, according to Kierkegaard, "the dizziness of freedom, which emerges when the spirit wants to posit the synthesis,[23] and freedom now looks down into its own possibility, laying hold of finiteness to support itself. In this dizziness, freedom collapses."[24]

Did Kierkegaard, with this psychology of *Angst,* achieve his purpose of making a first incidence of sin psychologically comprehensible, thus rescuing the biblical depiction of the innocent original state and a Fall that followed after it from Schleiermacher's criticism of its literal accuracy? That has to be regarded as doubtful, although, for example, Paul Tillich developed his doctrine of sin on the basis of Kierkegaard's writings.[25] Does not anxiety about oneself, the dizzying experience of the freedom that our self-awareness shows to be self-referential, not presuppose sin, which, in fact, consists of the human's being the center of his or her own world? Is, then, anxiety about oneself from the outset counter to a believing trust in God, an opposition that characterizes this anxiety as unbelief and so as sin? Martin Heidegger's interpretation of the phenomenon of *Angst* as paradigm for the fundamental structure of human existence as worry or concern (*Sorge*) points in this direction:

Anxiety, in fact, reveals that human beings, in their relationship to the world, are fundamentally concerned about themselves.[26] But that concern for oneself is the basic structure of human life is in itself an expression of the dominating role of self-love, the Augustinian *amor sui,* in human lives. Rudolf Bultmann, therefore, was right to judge worry to be the fundamental structural moment of the "pre-believing existence" of the human being as sinner;[27] for, to the extent that we worry about ourselves in the sense of Heidegger's cautious circumspection, we are no longer living in and by a trust that sustains our whole life but through a striving for assurance. This is not to say that such a concerned and anxious striving for assurance and security might somehow be simply avoidable and superfluous. It is not. But it implies the danger of falling short of oneself. And the more we surrender ourselves to the striving for security and control of the conditions of our life, the more is our life ruled by *amor sui* in Augustine's sense. In that Heidegger discovered, in the phenomenon of *Angst,* the basic structure of human existence as concern, his analysis thus implicitly confirmed that *Angst* is an expression of sin because it is an expression of human beings' concern for themselves. But that means, as regards the evaluation of Kierkegaard's investigation of the concept of *Angst,* that it has its enduring significance not in the function intended by Kierkegaard—in a psychology of the origins of sin—but as a description of the effects of sin on human self-awareness.

This conclusion has immediate consequences for the theological evaluation of aggression: Not only forms of aggression that are traceable to the motive of envy but also forms proceeding from anxiety and frustration are to be seen as expressions of that *fundamental failure to attain the full form of human existence* that is described by the theological term *sin.* We must immediately add that this characteristic does *not* identify aggressive individuals as a group of evildoers separated and to be separated from the rest of society. For the personal failure described by the concept of sin has a general human character. It is simply that, for entirely secondary reasons, it expresses itself differently in those who are manifestly aggressive than it does in the rest of society. The description of aggression, thus, stands in opposition to the possible consequences of dualistic theories of drives that regard aggression as pure destructivity.

# Aggression and the Theological Doctrine of Sin

Ordering it within the context of the theology of sin means to presume of aggressive persons as of other people that they are beloved by God and destined to be God's images, but, for what are ultimately reasons of structural shaping of behavior that are relevant to all human beings, they have fallen short of their human destiny and suffer from that failure of the self. On the other hand, the theological interpretation of aggression here presented also stands against tendencies to excuse aggressive behavior, such as those following Alfred Adler's association of aggression with the drive for self-preservation and especially those that have been employed in connection with the frustration theory of aggression.[28] If aggression is, in substance, only a human striving, justified in principle, toward self-preservation and self-development that, in aggressive individuals, has been frustrated so that these individuals are only responding with aggression, we are close to playing down and excusing aggression. It can then seem as if criticism should be directed at the cause of the frustration, not at the reaction to it. Helmut Nolte, himself an advocate of the frustration theory, saw this problematic tendency: "[J]ustifying aggression in the service of self-assertion and self-development unwittingly shifts to a justification of destruction. . . ."[29] But Nolte himself contributes to playing down and excusing aggression when he attributes to the capacity for aggression the positive function of "*supporting* the development and accomplishment of the individualizing achievements of self-preservation and self-realization . . . and when it encounters external resistance [it becomes] destructive, in accordance with inborn, physiologically grounded patterns of reaction . . . when necessary, against one's own person."[30] In the background stands the all-too-mechanical model of drives, according to which the aggressions built up by frustration must in every case be "released" in one way or another. This mechanism has been described, somewhat inaccurately, in terms of the concept derived from the aesthetic theory of the ancient tragedies, that of catharsis. This emerges, for example, from the psychology of learning's critique of the assumption of soothing the aggressive drive by satisfying it: The expression of aggression can very well lead to broader and intensified aggression, both in the social context, whose climate is heated by aggressive actions and in the aggressive individual himself or herself. Contrariwise, the renunciation of aggressive action can ease the renewed socialization of the one who is frustrated or otherwise inclined to aggression. Catharsis—purification, in the original sense of the word—consists precisely in overcoming aggression by release from the narrow bonds of self-absorption that underlie aggressive behavior and by being elevated to a common standpoint in the social life of the community in which the individual can find his or her acknowledged place.

Thus, finally, comes the question of what a theological interpretation and classification of aggression can contribute to overcoming or restraining it. We certainly should not have any exaggerated expectations of what theoretical theological constructions can effect. But there may be a contribution, not to be underestimated, in addressing, on the one hand, the demonizing and, on the other hand, the playing down and exculpating justification of aggression. The demonizing of aggression conjures up forms of its suppression that are in themselves aggressive. The playing down of aggression and the decontrol of aggressive expressions, however, seem to lead not so much to any satisfaction of a specific aggressive drive as to an *increase* of aggression in the overall social situation.

The question of what effectiveness promising strategies may offer society in preventing aggressive forms of behavior from running rampant, and thus also in protecting the members of society, is not something primarily for theology to answer; it falls primarily within the purview of other disciplines. What theology can do first of all is to contribute to removing factors within Christianity itself that favor aggression. These include the self-aggression bound up with the doctrine of original sin, which has not only—as discussed above—distorted Christianity's attitude toward sexuality but, beyond that, has promoted a climate of false penitence extending to the fear of hell and purgatory. Among further factors favoring the rise of aggression in Christian piety, we should also list "the constant excessive demands of the maximal commandments in Christian ethics taught by early Christian groups with a radical-ascetic piety in a very different situation."[31] Such maximal demands necessarily create frustrations, especially when they remain abstract and have no meaningful and effectual application to the concrete realities of life. The image of the punishing God is intimately connected to this intensification of ethical demands and the cul-

tivation of guilt feelings,[32] which can act as a source of self-aggression
and also, in the sense intended by the psychology of learning, as an
example of aggression directed outward. Here theology can usefully
reflect on the connection between deeds and consequences, founded in
the nature of things, which in the Old Testament underlies the idea of
divine punishments,[33] which thus are not the expression of arbitrary
vindictive power. The notion of divine punishment here functions only
to relate the complex of act and consequences to divine omnipotence,
so that it does not remain a lawless sphere of fatal interweaving of acts
and consequences, independent of God's action. Nevertheless, this view
of the internal connection between deeds and consequences is the starting point for the effort to relate it to divine omnipotence. The same
notion still underlies Paul's understanding of the essential connection
between sin and death.

All the factors we have mentioned as promoting aggression favor
self-aggression first of all, but it can at any moment shift to aggression
against another, as has happened, in the history of Christianity, especially in connection with the intolerance that so easily arises from a
sense of having received an exclusive revealed truth. We know today
that this intolerant exclusivity is contrary to the knowledge, fundamental to Christian faith, of how far Christians are from the eschatological perfection of the reign of God. The awareness of that difference,
expressed in Jesus' obedience as Son to his Father and in Paul's theology of the cross, is the condition for the community with God of which
Christian faith is so proud.

The appearance of motifs that promote aggression in the history of
Christianity, indeed in Christian doctrine itself, is, in light of what has
been said here about the relationship between sin and aggression, to be
regarded as a sign of the effective working of sin even in the church,
despite the fact that the church's life derives from the overcoming of
sin through the redeeming action of God in Jesus Christ. That such
deviations in Christianity can be completely overcome as the result of
theological insights is not to be expected. At any rate, we will have
to anticipate the appearance of other forms of the same problem. But
such a sober assessment of the situation will certainly lead us to expect,
in light of the general structure of social life, that the phenomenon of

aggression is much too tightly interwoven with the overall structures of human behavior and their tendency to fall short of their true being for us to hope for anything more than a curbing of aggression. Nevertheless, the connection between anxiety and aggression teaches us that the more we succeed in overcoming anxiety through trust and the experience of a meaningful identity for individuals within the social context, the more can aggression be reduced at its root. Of course, anxiety is a multiform phenomenon, as multiform as the human being. It cannot be alleviated by externally orderly and comfortable conditions of life. Human anxiety about the self is decisively directed toward a meaningful life. The magnitude of this problem will not be underestimated by those who recall that overcoming *Angst* required the revelation of the love of God in the cross of Jesus Christ. That is why the Christ of the fourth gospel says: "In this world you have anxiety, but fear not: I have overcome the world" (John 16:33).

PART FOUR

# MEANING AND
# METAPHYSICS

## ✦ I 2 ✦

# Meaning, Religion, and the
# Question of God

A MEANINGFUL LIFE is no longer taken for granted in the modern world. The concern with emptiness and loss of meaning,
together with a questioning about and searching after meaning, has
become a predominant theme in our time. As early as 1925, Paul Tillich
suggested that the question of meaning has attained as fundamental a
significance for modern folk as the question of overcoming transitoriness had for people of antiquity and the striving for forgiveness from
sin had in the medieval world. For Tillich, all individual meaning is
dependent on an unconditioned "ground of meaning," which both surpasses and serves as foundation for the totality of all individual semantic (meaning-related) contents.[1] In a largely similar manner, Viktor
Frankl has spoken since the Second World War of an "unconditional
meaning" [*Über-Sinn*] that grounds that meaning of existence without
which humans could not exist.[2] Like Tillich, Frankl perceives clearly
that one is concerned here in the final analysis with the religious quest.
Nevertheless, in the experience of the "lack of meaning," that malady
of our times is visible which stems from our secular society's disregard
for God and which according to Frankl provides the explanation for
the dramatic rise in the number of neurotic illnesses and especially of
suicides.

All such inquiries into meaning are concerned with what it is to possess meaning, that is, with the possibility of a life that even in suffering could be experienced and affirmed as meaningful. The meaning-filled life cannot be presupposed or taken for granted, as the experiences of emptiness and meaninglessness demonstrate clearly enough. From this observation many assume that we must create our own meaning and thus impart meaning to a reality that appears meaningless. Indeed, this view is dominant in contemporary sociology of knowledge under the influence of Edmund Husserl, Alfred Schütz, and Theodor Lessing. Thus, for Peter Berger the human formation of culture is fundamentally a matter of the creation of meaning; similarly, Niklas Luhmann views "the overcoming of contingence" as the most foundational accomplishment of a social system. From here it is only a short step to viewing the overcoming of experienced meaninglessness also for the *individual* as a task which involves a human creation of meaning. Solving the problem would then only depend on finding the power to give meaning to one's own life, in order to extricate oneself from the crippling influence of the Medusan countenance of meaninglessness.

But is the experience of meaning a matter of creating meaning or of discovering an already given meaning? In order to pursue this question, it is necessary to distinguish a normal concept of meaning from that of the meaning-filled life. The formal concept is more comprehensive than the actually meaningful. For example, the experience of the *absence* of meaning is also semantically structured and thus not devoid of meaning; the same pertains as well to the nihilistic denial of a meaningful world. Indeed, it is only because of the semantic or meaning-related structure of language that one can even articulate the conviction of the meaninglessness of life.

The distinction of a formal notion of what it is to be meaningful from actual meaning-filled content (Gerhard Sauter) is suggested to us also by a study of what is contained in the sentences of a discourse or text and which is grasped linguistically. This type of meaning is concerned with the meaning of the words in a sentence and of the sentences in the context of a discourse. The individual words have their meaning not only as designations for objects and states of affairs but also through their positions in the sentence.

Thinkers have attempted to draw a neat distinction between the two concepts "meaning" [*Bedeutung*] and "sense" [*Sinn*]. For instance, Gottlob Frege spoke of the meaning of the words as names for objects, and opposed this concept to the sense of the sentence as a whole. In his view sense has to do with the whole within which the words are arranged as components of the sentence. Now, it may as a matter of fact be the case that the concept of "sense" does belong primarily to sentences and that of "meaning" to words. However, the words have their meanings initially within sentences, and this meaning is not completely separable from the context of an individual sentence. A sentence is not merely a mechanical construction of words whose meanings are already set. Rather, the individual word, taken alone, always bears a certain degree of indeterminacy. It is no coincidence that dictionaries offer various nuances of meaning for each word, nuances which are abstracted from the word's actual use in sentences.

In a sentence the individual word receives a higher degree of semantic determinacy. This is because in a sentence the word bears meaning in a second sense, namely, as a constituent of the sentence. Here we normally speak of the *sense* of the word within the context of the sentence. It is not only the sentence as a whole that has a sense but also its individual constituents, the individual words: In the words the sense of the sentence is articulated. Thus, sense and meaning belong together; they resist a neat assignment to sentences and words. It is especially important, though, to differentiate two aspects within the notion of the word "meaning" itself: the reference to an object, and the position of the individual word in the sentence. Since meaning has to do with the position of particulars within the context of the whole, it is thus possible to speak also of the meaning of the particular sentence within the broader context of a discourse or text.

Linguistic meaning has therefore to do with the relationships between parts and a whole within the context of a discourse. At the same time, however, we are concerned with the subject which is being spoken about and which is "represented" through the mediation of the meanings of the words that make up the sentence. Now, of course, language has not only a representational function but expressive and communicative functions as well. There are forms of linguistic expression in which these other functions occupy the foreground. Nonetheless, the representational function always plays a part and may in turn move to the fore, namely, in the case of assertorial sentences. Assertions claim to be true in the sense that the meaning of such sentences attempts to represent an objectively existing meaning, a state of affairs. It is this truth claim which constitutes the sense or import of such sentences *qua* assertions.

Does the sense that linguistic utterances have owe its existence to a human bestowal of meaning? At first glance, this appears indeed to be the case. Sentences are, for example, spoken *by us,* leading us to think that their meaning is the result of our efforts. Since meaning can only be grasped linguistically, the belief that language is the product of human activity suggests to us that we might view all meaning as the product of a human bestowal of meaning. However, if we do so, two factors that are crucial to the semantic structure of linguistic utterances drop from sight.

In the first place, this view fails to consider that it is part of the nature of language itself to represent a reality that is already given, as we saw in our examination of the assertion. Even if only a few assertions are "true," it cannot be the case that all asserted meaning is only the expression of a human bestowal of meaning. True assertions are true precisely in that their content corresponds to the state of affairs that is being asserted. Now, the spoken or written sentence may be the product of a human activity as well; nonetheless, true sentences and true assertions are related to the reality of the asserted state of affairs in the sense of a discovery of meaning rather than in the sense of a bestowal of meaning.

A second important factor is the fact that there are many layers to the meaning of linguistic utterances. A spoken sentence always brings to expression something above and beyond the meaning that the speaker supposed or intended. It is not unusual that we say in reality something different from what we wanted to say. This is only possible insofar as the meaning of a sentence, once it has been spoken, proceeds from the combination of the words themselves, independently of the intentions which the speaker had in speaking it. A sentence can say more than the speaker actually wanted to say. It can also fall short of the thought which she wanted to express and which can be independently inferred from the context of her speech. Finally, a sentence can convey something completely different from what she intended. All of these things are matters of the interpretation of what was said. Moreover, every linguistic expression stands in need of interpretation by the listener or reader.

Nevertheless, interpretations can miss the meaning which the author intended the utterances to have, as well as the meaning which actually should have been derived from what was said. This possibility of error weighs heavily against the view that interpretations are only a bestowal of meaning. If the interpretation can miss the meaning of its object, then the meaning of a sentence, a discourse, or a text is obviously not merely dependent on the interpreter. Nevertheless, as we saw, meaning does not depend only on the speaker or author of the text. For these reasons, the semantic structure of the texts that we interpret appears to be an independent entity, and the appropriateness or inappropriateness of interpretations must be judged in relation to it.

In a similar manner, assertions also presuppose rather than produce the meaning of the corresponding state of affairs. Assertorial sentences rely unavoidably upon the meaning structures of states of affairs, which are prior to human perception and its articulation in language. Meaning can be approached through language, but it is not the product of language. Otherwise, all speaking with assertorial sentences would be misguided. If the use of assertions is meaningful—that is, if they express the particular nature of human experience and experienced reality—then reality must be somehow meaningfully structured prior to its being grasped in language, even if language is the only way to articulate this meaning structure. Language can either grasp or miss the semantic structure of reality, and therefore this semantic structure is not first created through language. To reduce meaning to language is to take the first step along a path which culminates in the position that all meaning is merely created through human action—that is, that it is the product of a bestowal of meaning.

Yet this position falsifies the actual state of affairs by reversing the actual priority. In fact, human action is itself dependent on perceptions of meaning, since there is no action without goal setting, which requires the choice of the means relevant to a given goal. This process always presupposes an orientation to the world and the grasp of semantic content. Now, a perceptive grasp of semantic content is admittedly itself an activity [*Tätigkeit*], but it is not an action [*Handeln*]; it does not realize self-chosen ends through the use of means. Nor is speaking and the grasp of reality through language—contrary to the prejudgments of contemporary speech act theories—always an action. We form verbal utterances only secondarily, as means for the attainment of selected goals or as ends in themselves; in both these special cases they are moments of an action. But that is not the fundamental character of spoken language.

This criticism of the reduction of linguistically grasped meaning to acts of human meaning bestowal is of fundamental significance for our theme. The connection of religion and the experience of meaning can only be conceptualized appropriately if the experienced meaning is seen to precede its comprehension by humans rather than being understood solely as the product of a human bestowal of meaning. If the latter were the case, religion would be merely a human projection, lacking any truth content that surpasses the human consciousness. But we have seen that the reduction of the perception and comprehension of meaning to a bestowal of meaning pulls the ground out from under the very notion of the truth of assertions themselves, not only from the truth-claims of religious statements.

The connection between religion and the experience of meaning becomes visible when one turns from the meaning-structures of linguistic utterances to the way in which human experiencing [*Erleben*] can be meaningful. Human experiencing is a special case of the semantic structure of reality itself, which we must consider as preceding its linguistic representation. Experiencing has to do with the ontological structure of a creature who is capable of language, and thus with the social context within which language is developed and used. In this phenomenon, the foundation of linguistically articulated meaning in prelinguistic meaning structure is perceptible in a special way.

We owe the first foundational analyses of the experience of meaning to Wilhelm Dilthey, who dealt with the semantic structures of experiencing in his late notes and sketches, which in turn formed the starting point for Martin Heidegger's analyses of individual human existence [*Dasein*]. Dilthey transformed the discussion of words in context to an inquiry into the structure of experiencing. He did not explicitly discuss this transition, since he presupposed that the meaning structures found in language themselves were only the expression of the meaning relatedness of the psychic life. For this reason Dilthey believed that it was possible to speak about meaning and meaning relationships only in the realm of the *psyche*.

For us, however, the transition from linguistic meaning to the postulation of a semantic structure to prelinguistic reality requires a justification that can only be obtained through observations of language itself. In the foregoing argumentation, we have appealed to the representational function of language, and particularly to the structure of assertions, to justify the supposition of structures of meaning that extend beyond the realm of the linguistic. In so doing, we have defended a much more extensive acceptance of semantic structures than did Dilthey, who limited them to the realm of the psychic life. In contrast to his position, we may now expect, in *all* realms of reality, that particular appearances can be understood as parts of more complete meaning-forms [*Sinngestalten*]—that is, contexts of meaning in Dilthey's sense exist everywhere, even beyond the realm of the phenomena of organic and psychic life.

Although I have argued for a wider context, Dilthey's special case of the human life-context does carry particular significance for the perception of meaning, since for humans the whole of their lives is present at every moment along with the particulars of their own experiencing. Dilthey, at any rate, expressed it in this way with his concept of experiencing. An individual event becomes an experience to the extent that it is grasped as one specific articulation of a whole life. Perhaps Dilthey construed the notion of experience too narrowly by relating it only to the whole of the individual life.

Heidegger's analysis of *Dasein* in *Being and Time* suffers from the same shortcoming. We have no specific consciousness of the whole of our own life (in contradistinction to all else) at the moment of immediate experiencing. Much more, it is the whole of reality itself that is present to us in feeling, not only the whole of our own life. In such a vague presence of reality itself, world, self, and God are as yet undifferentiated. The whole has definiteness only in the particular experience. The individual occasion of experiencing, though, is not just something particular; in it, the whole of reality appears—just as the meaning-context of a discourse appears in the individual words and sentences. In experiencing, the whole of reality is not fully contained in the individual experience; there remains a vague element of "above and beyond," which at the same time forms the framework in which the individual experience can first become what it is. There is—as modern philosophy since René Descartes has seen, and as medieval Scholastic thought already knew—a vague awareness of an undetermined infinite which always precedes all comprehension of anything finite or determined. As Descartes said, the finite can only be comprehended as a limitation of this infinite.

This, then, is the background of Dilthey's concept of experiencing. Dilthey may have identified himself more with Friedrich Schleiermacher and Benedictus de Spinoza than with Descartes, but in this matter Spinoza was only a student of Descartes. Dilthey narrowed the horizon of the undetermined infinite and whole, which is present to us in our affections as the horizon of our individual experiences, to the totality of life—indeed, of the individual's own life. He gained thereby the basis for centering his philosophy on the concept of life [*Lebensphilosophie*] and for his descriptions of the ontological structure of experiencing, as well as for his view of the human experience of the self as a process of self-interpretation. Under this view, as a life history progresses the meaning structures of earlier experiences shift, because the whole of life appears again and again under new perspectives, that is, from the viewpoint of new experiences. What was earlier experienced as important becomes unimportant, and other scarcely noticed moments of earlier experiencing can increase in significance. Thus Dilthey writes: "Not until the last moment of a life can the final estimate of its meaning be made."[3] Until then, the meaning of the particular moments of experiencing shifts, as does the meaning of the whole of one's life. Herein lies the finitude of our knowledge about life as a whole. We can never attain a comprehensive overview of the total meaning of our life—not because we have absolutely no relationship to our life as a whole, but because we always have such a relationship to our life and to life in general only from the limited viewpoint of a specific experience, from which we remember earlier experiences and await future ones. This viewpoint changes as our personal history progresses through time. Consequently, we possess the whole, the total meaning of life, only in the manner in which it is represented in the respective individual experiences.

It is amazing to note how closely this description of the semantic structure of experiencing in Dilthey's thought is connected with Schleiermacher's description of religious experience in the second of his *Speeches* (1799). In that work religious experience is an intuition [*Anschauung*] of the infinite and whole in one individual, finite content. We come to such a view of the universe when we become cognizant that what is individual and finite does not exist for itself but rather is "cut out," together with its boundaries which constitute its particularity, of the infinite and the whole. In point of fact, this is the same conception which we can find already in Descartes, that we can only comprehend finite objects through a circumscription of the infinite. We are, however—as Schleiermacher further points out—normally not aware of this fact in our everyday lives, interacting as we do with finite objects and states of affairs as if they had their existence from and in themselves. It is only in the higher awareness of religious experiences that we become aware of the actual, deeper reality of things, namely, that they are constituted by and through the "universe," that is, the infinite and the whole. According to Schleiermacher, this higher perceptual awareness constitutes the unique essence of religious experience. Yet even religious awareness can grasp the universe only through intuitions of finite things and states of affairs. For this reason, Schleiermacher allows for an indefinite multiplicity of various forms of religious consciousness according to the character of the particular intuition by means of which the universe is comprehended, since each intuition is seen as a part of the whole and thus as a revelation of the whole.

Here is the point of contact with Dilthey. We "have" the whole of life, its total meaning, only in the individual and the specific, in which the whole manifests itself. Now, human experience can attain unity in the midst of the multiplicity of impressions and intuitions by means of a ruling or guiding intuition to which everything else is related. But even when one such integrating intuition becomes dominant—a phenomenon which, according to Schleiermacher, underlies individual religious life histories as well as the origin and development of the historical religions—this intuition still remains bound to a particular viewpoint, in a manner similar to Dilthey's position concerning the experience or significance of one's own life.

Perhaps Dilthey highlighted the historicity of this process more strongly than Schleiermacher. For example, he speaks of the possibility of a final knowledge of the meaning of our existence at the end of our life (or, in the case of the history of humanity, at the end of history itself). However, this eschatological possibility of a final decision about the meaning of life and of its individual moments must actually lie at some point beyond life, since a life is over at its "last moment." For Dilthey, occasional comments like these function only to underscore the point that a final knowledge of the total meaning of life is inaccessible—and in this matter he is in complete agreement with Schleiermacher.

Of course, Dilthey's closeness to Schleiermacher here is not coincidental: He occupied himself intensely with Schleiermacher's thought for decades, and a major biography of Schleiermacher belongs among Dilthey's chief works. It is thus not surprising that Dilthey was also influenced by Schleiermacher in his systematic thought. As we saw, he followed Schleiermacher's understanding of the finiteness and particularity of our awareness of the whole of life. In opposition to G. W. F. Hegel, who believed it possible to comprehend the infinite whole unbrokenly in the form of Idea, Dilthey also appropriated Schleiermacher's view of the relativity of this awareness to specific experiences within one's own history.

This awareness—that we have the whole only in and through the fragments—links Dilthey to Schleiermacher. Yet for Dilthey the whole is conceptualized differently than for Schleiermacher: it no longer connotes the universe of reality itself, but rather the whole of "life" in the process of its history. The shift has its basis in Dilthey's limitation of the concept of meaning to life, as we saw above. Tied to this limitation is the fact that Dilthey no longer spoke in an explicitly religious manner of the presence of the whole in experiencing but only related this state of affairs to the theme of the experience of the self.

I have dealt with Dilthey in such detail because of the incisive nature of his analyses of the semantic structure of human experiencing, analyses which are fundamental for the contemporary discussion. This is especially true of his position on the significance of individual moments in the context of the whole, a whole which always remains incomplete for the experiencing individuals themselves during the process of their history. Wherever the question of meaning is related to the whole of life and of experienced reality—as, for example, in Tillich's work—such that each individual meaning possesses its significance only from an all-encompassing context of meaning, there Dilthey's analyses can be detected in the background. In Frankl's psychology we also meet up with this understanding of meaning, which is occupied with the relationship of parts of life to the whole and with the presence of this whole in the individual experience. In contrast to Dilthey, Frankl seeks in this way to do more than merely describe the meaning-structure of experiencing. Whereas such a description would leave open the question of whether life is actually experienced as meaningful or as meaningless, Frankl also desires to encourage trust in life's meaningfulness through a total meaning [*Gesamtsinn*] which encompasses life as a whole, though for him such a total meaning can only be grasped indirectly through the mediation of, and in, concrete life-situations. Once again the semantic structure of experiencing, viewed formally, proves to be linked with the religious theme.

Before we pursue the question of the particularity of the religious awareness of meaning in its relationship to the semantic structuredness of human experiencing generally, we should first emphasize a point which represents perhaps the most important gain provided by Dilthey's analyses of meaning and significance in the context of experiencing. Dilthey's descriptions offer an understanding of meaning and significance according to which these do not stem from a bestowal of meaning by the human subject but proceed from the relationships of life itself, that is, from the relationships of its submoments to the whole of the life-context. Viewed in this way, events already have meaning and significance. This applies also to the events of history, which do not need to have a meaning subsequently conferred upon them through human interpretation. Historical events have meaning and significance themselves according to their contribution to the whole of the life context in which they belong. To be sure, the meaning and significance of the individual events can be determined only relative to the standpoint of historical consciousness.

Dilthey was able, then, through reflection on the historicity of the historical consciousness itself, to do justice to the multiplicity of interpretations of historical occurrences, as well as to the significance which accrues to each but which cannot be fully determined until the end of history. Life's moments have a significance [*Bedeutung*] in themselves, but we can only grasp their significance through the medium of an interpretation [*Deutung*] which itself is conditioned by the perspective of a particular historical standpoint. This insight is valid for the life experience of the individual just as much as for history at large. Only from the end of history could we fully and completely comprehend the significance inherent in the events and forms of history. Only from the end of history, therefore, will a final decision concerning the truth or falsity of our convictions of meaning be made. The evidence which the contemporary experience of meaning provides has the form of faith and of an anticipatory representation of a meaning which has yet to appear with finality.

The relationship between Dilthey's description of the experience of meaning in everyday contexts on the one hand, and the specifically religious consciousness of meaning on the other, has been in principle already elucidated by Schleiermacher. He described the everyday consciousness as oriented toward finite objects and relationships, whereas the religious consciousness comprehends finite realities as grounded in the infinite and whole, thereby intuiting the infinite itself in the finite things. In 1925 Paul Tillich wrote that all individual meaning is conditioned by a context of meaning, which in turn rests on an unconditioned ground of meaning. This unconditioned ground of meaning, however, only becomes an explicit topic for the religious consciousness. The cultural consciousness, which is oriented around individual meaning, presupposes such an unconditioned meaning but does not occupy itself expressly with it: "Every cultural act contains the unconditioned meaning; it is based upon the ground of meaning; insofar as it is an act of meaning it is substantially religious." But it is not expressly religious: "Religion is directedness toward the Unconditional, and culture is directedness toward the conditioned forms and their unity."[4]

I gave expression to a similar determination of the relationship of the religious consciousness to the semantic structure of everyday experience in *Theology and the Philosophy of Science*, linking myself more closely than Tillich, however, to the hermeneutical analyses of meaning propounded by Dilthey.[5] I argued there that the religious consciousness has as its explicit theme that totality of meaning which is implicitly presupposed in all everyday experiences of meaning, oriented as they are around individual experiences of significance. Religion has above all to do with the divine reality that grounds and completes the meaning totality of the natural and social world, and thus only indirectly with the totality of meaning of the world itself. Nevertheless, the truth-claim made by the religious consciousness must authenticate itself by showing that the God (or gods) alleged by it can actually be understood as the creator and perfecter of the world as in fact experienced. The assertions made by the religious traditions, which are directed beyond formal meaningfulness to the quest for meaning in human life, must prove themselves: they must be able to integrate the relations implicit in everyday experiences of meaning within an encompassing context of meaning that grounds all individual meaning. The urgent experiences of senselessness, suffering, and evil are among those life experiences which the religious consciousness of meaning must integrate. If a specific religious tradition is not able to do justice to human experience through such integration, its failure will lead to a crisis of belief in the truth of the tradition; it then becomes questionable whether the God proclaimed by this tradition can, as a matter of fact, be understood as and believed to be the creator and perfecter of the world as actually experienced by humans.

Christian truth-claims about God must also face this question of a confirmation through the human experience of meaning and its implications for the understanding of reality as a whole. The feelings, so widespread today, of an all-pervading senselessness, together with the related questioning after meaning, indicate that, for many persons and for broad segments of the public consciousness in our secular culture, the traditional answers of Christianity are no longer adequately functioning as a comprehensive interpretation of the experience of the world's reality and of the life problems that contemporary people face. The individual reasons for this failure cannot be developed here. However, the contemporary question of meaning that arises out of the experience of the absence of meaning should not simply be dismissed by Christian theology as an idolatrous question.[6] Certainly, theology must criticize the widespread tendency to reduce meaning to human action as self-destructive.[7] It is also correct that meaning and truth are not the same.[8] Seductive images may be experienced as most meaningful—that is the key to their seductiveness, since only for this reason can they lead astray.

Attention to the suffering of meaninglessness can create the false impression that the problem might be solved simply by providing humans with some sort of sense that life is meaningful, as if the content were a peripheral matter and the question of the truth or falsity superfluous and disruptive.[9] Yet if we were to approach the question of meaning that arises out of the experience of meaninglessness as if it were merely a demand to anesthetize nihilistic experience, we would have misunderstood it. Those who earnestly inquire into meaning are concerned with an adequate answer to the problems which have led to the forfeiture of the consciousness of meaning.

Thus the question of meaning, correctly understood, is inseparable from the question of truth. This is evidenced by the longing for an all-encompassing meaning. For to the concept of truth belongs the unity of all truth, that is, the simultaneous existence, without contradiction, of each individual truth with all other truths. From this insight alone it should be clear that the question of the meaning-context of reality as a whole is not theologically illegitimate.[10] To inquire into the total meaning of reality is not automatically an expression of human presumptuousness. It is a matter of fact that the individual is everywhere conditioned by the whole, and the consciousness of this state of affairs belongs essentially to what it means to be human. To be sure, the simultaneous awareness that we can never gain a definitive overview of the whole of reality is also a part of our humanity. Only when this is forgotten is it appropriate to speak of presumptuousness.

# Meaning, Religion, and the Question of God

Knowledge about the whole of reality itself and the question of its basis must not be confused with this sort of presumptuousness. The presumption lies in alleging to command a definitive view over the whole, whereby persons forget their own finitude and place themselves in the position of God. In contrast, the sort of knowledge of the whole of reality that remains conscious at the same time of its own finiteness reaches consummation in a knowledge of God as distinct from human subjectivity. The idea of God as such is always an answer to the question of the meaning of reality as a whole. Whoever wishes to exclude this question must also forbid that religious consciousness through which we honor God as the creator of ourselves and the world.

It is certainly appropriate to a correct knowledge of God that we include an allusion to divine inscrutability. Yet this allusion must not be understood as an attempt to avoid answering the question of the meaning-context of reality as a whole. Instead, it represents a phase in such an answer, insofar as it emphasizes the superiority of the God-based meaning of the life-world as a whole over and above the limitations of human understanding. Even negative theology, which refuses to go beyond the conception of the unknown God, is in this sense an answer to the human question of meaning. Of course, it is not the Christian answer, for Christianity confesses that in Jesus of Nazareth the divine *Logos* has become human, the one in whom all things have their being. The Greek word *logos* connotes "meaning" as much as "word." The connection of the Old Testament concept of the divine Word with the Greek notion of *logos* means nothing less than that the context of meaning which encompasses the entire creation and its history up through the eschatological completion has been made manifest in Jesus Christ.

# $\div$ I 3 $\div$

# Eternity, Time, and Space

THE CONCEPTS of space and time are not only important in physics and in geometry. They are of basic importance in all human
experience. It is not self-evident that the definition of these concepts is
an exclusive prerogative of geometry and physics. Certainly, the measurement of spatial and temporal relations is a matter of special competence of geometers and physicists. But it is by no means certain that the
measurement of spatial and temporal relations exhausts the concepts of
space and time.

In ordinary human experience, space is the order of togetherness
of simultaneous phenomena, time, the order of their sequence. It has
often been said, with good reason, that time is the more fundamental of the two because the concept of space as order of togetherness
presupposes already the temporal notion of simultaneity. Space is the
order of togetherness of simultaneously existing phenomena, especially
of physical bodies. Therefore, when the theory of relativity told us that
there is no exact simultaneity, this had incisive consequences for the
concept of space. Our concepts of space in distinction from time have
become approximations of what more accurately is described as space-
time. Still, the distinction of space from time retains its importance in
human experience, and in reverse certain conditions of space and time
also apply to the concept of space-time.

One such condition was emphasized by Immanuel Kant in his analysis of our concepts of space and time in his *Critique of Pure Reason* (1781). Kant argued that all our conceptions of spaces and of specific times presuppose a prior intuition of space as an infinite whole and of time as an infinite whole because different spaces can be distinguished and related only within some prior space and because different times are conceived as parts of one and the same time. Spatial division and composition can occur only within some comprehensive space. Similarly with time: It is only within time that temporal distinctions are possible. Therefore, some infinite and undivided whole of space and time is a prior condition in forming any conception of spatial or temporal units. This has an important consequence: The concepts of time and space as infinite wholes are prior to all geometry because the spatial and temporal units that are needed for measurement are themselves parts of space and of time, which precede as infinite wholes all notions of partial spaces and times. At this point, it should be evident that the description of space does not belong exclusively to geometry but is also and even primarily a matter of philosophy.

In the case of Kant, the thesis that space and time as infinite wholes are required as preconditions in conceiving of any partial spaces or times belongs to his analysis of human consciousness. However, the validity of the thesis is not restricted to human consciousness of space and time but applies to the objective content of experience as well. Kant took his argument from Samuel Clarke's correspondence with Leibniz: When defending Newton's notion of space and of its relation to the Deity in expressing the omnipresence of the Creator with his creatures, Clarke argued that, in all spatial division and composition, some infinite and undivided space is presupposed a space, within which operations of division and composition become possible. Clarke considered his infinite and undivided space to be the space of God's omnipresence and the effect of God's infinity in his relationship with the world of his creatures. By insisting on the priority of infinite and undivided space in all perception of spaces, Clarke met the criticism of Gottfried Wilhelm Leibniz, who had objected to Newton's theological interpretation of absolute space as an organon of God's presence with his creation.[1] Leibniz had argued that, on such an assumption, God would have to be composed of parts and divisible into parts. Clarke's response was that, no, God is neither composed of parts nor divisible into parts since the infinite space of his omnipresence is undivided, prior to all division and composition. The point of this argument was that Newton's theological interpretation of space in terms of *sensorium* or organon of God's presence with his creatures did not involve any pantheistic implications. But Clarke's insistence that the space of God's omnipresence was not only infinite but also undivided made it difficult to identify this space with Newton's own concept of absolute space because that absolute space had to have a metrical structure in order to guarantee the concept of straight lines involved in Newton's principle of inertia, according to which bodies tend to continue their movement in straight line unless disturbed by other forces.

From a modem perspective, the difference of Clarke's infinite space as undivided from Newton's concept of absolute space saves Clarke's argument from falling prey to the abolition of the idea of absolute space by the theory of general relativity. Even the relativistic concept of space-time works with measurement which needs units of measurement that are conceivable only within some prior, infinite, and undivided space, according to Samuel Clarke and Immanuel Kant. Therefore, the space-time concept of the relativity theory does not unsettle this basic philosophical analysis of space and time. Although relativity has an impact on the philosophical issues of space and time, as will be argued a little later, still it does not completely reconstitute our notions of space and time. William Lane Craig in a series of recent books, especially in his book *Time and Eternity: Exploring God's Relationship to Time* (2001), remarks, "At best, scientific accounts describe our *measures* of time, but not time itself,"[2] and furthermore, "Curved space-time is just a geometrical model of gravity."[3] But how is Samuel Clarke's theological interpretation of infinite space and time to be evaluated? Immanuel Kant already struggled with this question. Ten years before his *Critique of Pure Reason* was published, Kant still shared Clarke's theological interpretation of the infinite and undivided whole of space and time that is presupposed in all our spatial and temporal perceptions. In his dissertation *De Mundi Sensibilis atque Intelligibilis Forma et Principiis* (1770), Kant affirmed that the infinite and undivided space as condition of all spatial conception is the form in which the divine omnipresence appears in the world: *spatium dici potest omnipraesentia . . . Phaenom-enon*. Similarly, the infinite whole of time was said to express the divine eternity in its relation to the world: *conceptus temporis tamquam unici infiniti et immutabilis, in quo sunt et durant omnia, est causae genera-lis aeternitas phenomenon* (§22). Ten years later, however, in Kant's *Critique of Pure Reason*, this theological interpretation of space and time was silently eliminated. Instead, the unity of infinite space and time was now reconceived as based upon the unity of the human subject of experience, though it remained unclear how the human subject, which is finite, can account for the *objective validity* of our conception of the *infinite unity* of space and time that is presupposed in all experience. The change of Kant's thought on the issue of theological implications of the concepts of infinite time and space has been explained as a result of Kant's concern for God's transcendence regarding the world (H. Heim-soeth). The assumption that the infinite unity of time and space that is presupposed in all human experience expresses the divine eternity and omnipresence could indeed result in a pantheistic conception of God's immanence in the world. This consequence, however, could only occur if that infinite space would be identified with the space of Euclidean geometry and with Newton's absolute space. Kant could have protected himself against such a consequence by insisting, as Samuel Clarke had done, on the *undivided* nature of the space of God's omnipresence. In this case, he would have been left with the problem of how that undivided infinite space is related to Newton's concept of absolute space as a receptacle or container of things. Kant tried to avoid the idea of space as an infinite and empty receptacle of things, and so he opted for the alternative idea of Leibniz that space is a system of relations in the mind but, according to Kant, no longer in the mind of God but in the human subject of experience, the transcendental ego. The difficulty with this position is, as I already mentioned, how the human subject, which is finite, can guarantee the objective unity of the spatial and temporal world we experience.

Kant did not have at his disposal an alternative that is available since Einstein's theory of relativity. That is the concept of space-time, which does not only integrate the metrical systems of space and time but also the concepts of mass and energy, since the metrical structure of space and time is no longer conceived in abstraction from the presence of physical objects.[4] Rather, those physical objects are accounted for as effects of the gravitational field of space-time. In his preface to the book of Max Jammer on the concept of space, Albert Einstein emphasized in 1953 the importance of the field concept in replacing the fundamental role of the concept of physical bodies in physics and eliminating at the same time the concept of space as an empty container of physical bodies.[5]

But how is the concept of space-time related to eternity? How could it have helped Kant to avoid pantheistic consequences of Clarke's theological interpretation of infinite space and time as expressing God's eternity and omnipresence in his relationship with his creation? The first step toward an answer to this question is to realize that space-time should not be seen in the line of Clarke's concept of infinite and undivided space (and time) as a comprehensive precondition of any discernment of particular spaces (and times).

Space-time as a geometrical concept describes the comprehensive field of all finite phenomena, especially of matter and masses. Thus, it is already distinct from Clarke's infinite and undivided space (and time), which is prior to all geometry and consequently to space-time also. It is because of the connection of the space-time concept with the occurrence of material phenomena, of masses, in the universe that I said that Kant could have been helped by the concept of space-time in his struggle for an unambiguous distinction between God and the world. He could have been more confident in aligning himself with Samuel Clarke's insistence on the *undivided* nature of the space of God's omnipresence if he could have distinguished the world of finite experience, the world of nature, from God's eternity and omnipresence in terms of the geometrical field of space-times. To be sure, there is also a possibility of a pantheistic interpretation of space-time itself, as the example of Albert Einstein with his sympathies for Spinoza shows. But space-time is not eternity. The geometric description of time in terms of a further dimension in addition to the three dimensions of Euclidean space may suggest a similarity of space-time to the concept of eternity, where everything is simultaneous. But this is only the effect of the spacialization of time, where the differences of tense—the distinctions between present, past, and future—are removed from the picture. In the eternal present, simultaneity is not bought at such a price of abstraction, but in the eternal possession of the whole of life, the distinctions of tense like other forms of differentiation are preserved.

A pantheistic view of space-time suggests itself only if the undivided nature of infinite space (and time) is not distinguished as it should be from all geometrical descriptions of space (and time). Therefore, the priority of undivided infinite space (and time) with regard to any specific conceptions of spatial and temporal units and as a condition of their possibility is so important. It also means that the space of God's omnipresence is not a container space. Ideas of God's omnipresence are inexplicable without some connection with the concept of space, but it has to be a concept of space and time that is different from geometrical space and time, prior to all measurement, if the distinction between God and the world is to be observed. The eternal God is present in his creation without becoming a component of the physical world with the exception, perhaps, of his Incarnation in one individual human person.

This relationship of transcendence and immanence may receive some deeper elucidation from a discussion of the concept of eternity in its relation to time. The concept of eternity is certainly opposed to the transience in the temporal succession of events. Therefore, it has often been assumed that eternity is completely opposed to time, a present that does not change (*nunc stans*) in contrast to our present that is continuously changing (*nunc fluens*). This was the Augustinian view of eternity that was bound up with the concept of divine immutability. There is a different view of eternity, however, that should not be confused with that timeless eternity. Here, I do not think of the idea of a life everlasting, since that notion is deeply ambiguous. If it only means a life that is going on without end but otherwise is similar to our present form of life, there is no idea of eternity at all but only of time without end. By contrast, the alternative to timeless eternity that I have in mind is bound up with the totality of life as presently experienced. It is a view that was developed in the Platonic tradition of thought like that of Augustine but somewhat earlier. It is the Plotinian idea of eternity as simultaneous presence and possession of the wholeness of life, an idea that Plotinus developed in his *Enneads* III, 7, 3: What in our experience is separated by the course of time, in the sequence of temporal events, is present all at once in eternity. This idea is echoed in the famous sentence of Boethius, from his *Consolation of Philosophy*, that eternity is the complete possession all at once of interminable life (V, 6, 4: *interminabilis vitae tota simul et perfecta possessio*). This idea of eternity should not be confused with timelessness because it does not exclude the notion of a sequence of events, provided that such a sequence is enjoyed simultaneously as a whole. Like in the case of timeless eternity, the idea of unchanging identity is included, but the reference to the wholeness of life allows for a plurality of events in that life, events that may form a sequence among themselves but are integrated in the wholeness of that life that is enjoyed as present in its wholeness and therefore not subject to change. This idea of eternity could be called omnitemporal since it comprehends the wholeness of life, but not in the sense of an everlasting process but rather as continuous presence of the whole of life. This concept of eternity corresponds to the infinite unity of time that, according to Kant, is presupposed in every distinct notion of particular times. But while the case of Kant's idea of infinite time was conceived as empty time, the idea of eternity comprises the differentiated fullness of life as simultaneously present. In application to the doctrine of God, this concept of eternity comprises not only the atemporal existence of God prior to the creation of the world, a priority that is causal, but not temporal, since the act of creation should be understood with Augustine as involving the creation of time itself. The concept of eternity as simultaneous possession of the fullness of life, which is otherwise divided in the sequence of events, also comprises the participation of the eternal God in the history of his creation, the divine economy which is finally to be consummated in the eschatological participation of creation in God's own eternal life. This is the destiny of all creation that the apostle Paul spoke about in Romans 8, which involves an element of judgment and of transformation on the part of the creatures, since the perishable cannot inherit the imperishable (1 Cor. 15:50), not without a profound transformation, that is, because "this perishable nature must put on the imperishable," as Paul says (1 Cor. 15:53).

Thus, the eternal God is both transcendent and immanent in the world of his creation. Temporal sequence is the appropriate form of existence in the case of finite entities, whose present is different from their past and different from their future. But this temporal existence is related to a future of participation in God's eternal life. And the eternal God is active in the history of his creatures by drawing them into that future from the first moment of their existence. Thus, the future of God, which is identical with his eternal present when it becomes the destiny of his creatures, is already the creative source of their existence. It is the source of the contingent existence of each creature, corresponding to the contingency of creation at large, but also the source of the definitive identity of each creature. If the divine eternity in the sense of simultaneous presence and possession of the wholeness of life is understood as eternity of the trinitarian God, whose identity allows for differentiation and self-differentiation, then it also allows for a world of creatures that are different from God as well as from one another, and yet exist in the orbit of God's omnipresence, and are destined to participate finally in God's eternity without loosing their finite nature and identity in difference from their Creator. They are destined to participate in God's eternity, in his eternal life, precisely by accepting and acknowledging their difference from the eternal God because such acceptance is the condition of having communion with that God.

Thus, the time of the creatures is not completely cut off from eternity. Rather, as Plotinus already noticed, the transition from one moment to the next one requires an encompassing unity in the process, a reminder of the unity of life that otherwise seems lost in the incessant perishing of each present into the past that is no more and in the face of a future that is not yet. Plotinus thought that the loss of the wholeness of life in the separation of present, past, and future is a result of a "fall" from the original wholeness of life which nevertheless continues to be present to some extent in the sequence of events. Therefore, in his view, the separation of time from eternity is not absolute. Later on, Kant contributed an argument to the same effect: Even the single moment of time would not be conceivable except for an awareness of time as a whole, for only within that encompassing whole one moment or part of time can be discerned from others. Thus, the unity of time as an infinite whole, which is conceived as realized in the concept of eternity, is somehow present in the flow of time. It is particularly present in the experience of duration, which is always colored by memory and anticipation as Augustine argued in his analysis of time in Book XI of his *Confessiones,* where the experience of duration in spite of the brokenness of the temporal process is illustrated by the famous example of how we experience the unity of a piece of music, a melody, an experience that would not be possible in our attention without the help of memory and anticipation. Such an experience of duration can be a reminder of eternity, the simultaneous presence and possession of the wholeness of life, although in our temporal experience such duration is always limited and gets interrupted.

The experience of duration as in the case of perceiving a melody is much closer to the concept of eternity than the mere fact of continuity in the process of time is, though even here, as Plotinus said, the eternal unity of life is still present in the background as condition of the cohesiveness in the sequence of time. Even in this case, there is a distant similarity with eternity since the temporal process, in spite of its constitution by succeeding events, may be conceived as a whole. It is a more distant similarity, however, because it is not experienced as a whole by any member of the process, like in the case of experiencing the unity of a melody while it is sung. Still, the analogy with a melody could be applied to the process of the universe, conceived as a "song of the universe," *carmen universitatis.*

The model of a temporal process perceived by an observer as a whole seems to apply also to the model of space-time. In the geometrical model of space-time the process of the universe is perceived as a quasi-simultaneous unity. As such, it appears to the eyes of the theorist, however, not within itself. It is a spacialization of the natural process that can be taken as an analogy to the way everything is present to the eternal God. The difference, however, is that in the case of God's knowledge of the world of his creation, as far as we can imagine it, the temporal differentiation between earlier and later, as well as the differences between past, present, and future—relative to each creature—are preserved. Time is not an illusion in the eyes of the Creator, to whom all things are present.

Time is not an illusion because the Creator wanted the independent existence of creatures, and therefore, he created time itself. Time
is a condition of the existence of finite entities. They exist each in their
own time and their own place. Except for the most primitive forms of
created existence, the creatures also enjoy some duration and, hence,
some form of permanence, though limited, an existence of their own
which is intended in the very act of creation. Creating something means
to bestow some degree of independent existence upon the creature.
Organic creatures enjoy a higher degree of such an independent existence when they organize their own life to preserve and nourish themselves in relation to their environment. This needs time. Thus, it is in
a given span of time that these creatures can organize their own being
and acquire a more differentiated form of existence. We can understand that this aim of creation is only obtainable in time. Afterwards,
after a creature has acquired a form of existence of its own, it may be
preserved in eternity, as it is promised in the Bible. But in order to be
obtained, time is a necessary requirement for the formation of finite
beings. Thus, time is not, as Plotinus thought, the result of a fall from
a primordial unity of life but a condition of the independent existence
of creatures, especially of their formation by self-organization. Therefore, in a Christian view of creation, time is created by the Creator as
a condition of a somewhat independent existence of his creatures. A
similar consideration applies to space. In order to preserve and develop
their own existence, creatures need some space into which to grow and
to relate to others, as well as time. While in God's eternity, simultaneity, the principle of space, and everlasting continuity are united, in the
world of the creatures, they get separated into space and time as conditions of their finite existence.

Space and time, then, in their distinction from eternity, are not
independent realities, as the container view of space suggested that
goes back to Aristotle. It was rejected by Nicene theology, as Thomas
Torrance showed,[6] but was adopted in Western medieval thought and
also by Newton in his concept of absolute space.[7] Newton's concept of
absolute space became the most influential model of a container view
of space in its combination with Euclidean geometry. Samuel Clarke's
idea of an infinite and undivided space as a prior condition for any

conception of partial spaces was a different matter. His undivided infinite space, if clearly distinguished from geometrical space, was not a conception of an infinite container of bodies but of God's dynamic omnipresence with his creatures. On the other hand, an absolute geometrical space is, in fact, an empty container of things, and this conception has been destroyed by the relativity theory. At this point, the theory of relativity told the philosophy of time and space an important lesson, the lesson that there is an interdependence between physical objects and the spatial and temporal dimensions of their existence. There is no measurable time and space without creatures. The geometric description of this connection by the concept of space-time may be only an approximation, if we consider the alternative interpretations of relativity by Einstein and Neo-Lorentzians,[8] but the insight into the interrelatedness of space and time with masses and energies will remain a lasting contribution to the understanding of the conditions of finite reality, even in the discourse of philosophers and theologians. This means that God created time and space when he created the world of finite entities as dimensions of their existence.

The last statement applies in any case to measurable time and space. But how does such a view of space and time relate to our earlier affirmation, with Samuel Clarke and Immanuel Kant, that the possibility of any conception of partial spaces or times requires as a prior condition the conception of an infinite and undivided whole of space and time? Is this infinite and undivided whole of space and time, which is prior to all geometrical description, also a property that belongs to the world of finite entities, or is it—as Clarke assumed and Kant also believed until 1770—an effect of God's eternity and omnipresence with his creatures? While measurable space and time seem to belong to the finite entities that exist in space and time, the undivided infinite space and time that is prior to them seems closer to the concept of eternity, which also involves simultaneity and, therefore, omnipresence as soon as creatures are called into existence. Their finite existence involves their setting in measurable time and space, which is created with them but takes place within some more comprehensive, infinite, and undivided space and time. It takes place within the orbit of God's eternity and omnipresence, which is not, however, to be mistaken for an infinite container because that could not be without divisibility. Rather, God's eternity and omni-presence are the medium of God's powerful presence with his creatures at the place and time of their existence. In his eternity, then, God is transcendent as well as immanent regarding the world of his creation. The creatures exist in their measurable time and space and in the universe of space-times within the presence of the eternal God who infinitely transcends them and yet is not far from any of them.

# ⇥ 14 ⇤

## Atomism, Duration, Form
### *Difficulties with Process Philosophy*

W HOEVER COMES in contact with process philosophy today
encounters it more often than not in the form of Alfred North
Whitehead's philosophy. This was the way it was for me when I was
guest professor in 1963 at the University of Chicago and ran into an
entire school of Whitehead adherents in the theological faculty (not, I
must say, in the philosophical faculty). Consequently, for the sake of
my own intellectual survival, I had to come to grips quickly and inten-
sively with the writings of this philosopher, who was at that time hardly
known on the continent of Europe.

I found the experience enriching; it made up for a certain lack in
the great tradition of German Idealism. One feels in Idealism the need
for a philosophy of nature adequate to the demands of our century,
for a metaphysics that, from the outset, integrates the awareness of the
world as mediated by contemporary knowledge of nature with human
experience as disclosed by the humanities and social sciences. The
more time I spent with Whitehead, however, the more I was shocked
by the rather dogmatic way in which he is read in the United States
within the school of process theology, a school which has in the mean-
time become quite influential. In that school Whitehead is taken to be
an entirely self-sufficient systematic thinker and, as such, authoritative,

like Aristotle in the High Scholasticism of the thirteenth century. He is not read as an exponent of a broader current of process thought, in the context of which his philosophical approach represents only one of the options which can be (and in part already have been) pursued.

If one considers Whitehead's philosophy in its relationship to thinkers such as Henri Bergson and Samuel Alexander (to name only two), then one becomes aware of different versions of the process-philosophical perspective. It becomes clear that a process-philosophical approach—one which dismisses the notion of a timelessly identical substance—is not necessarily tied to the specific assumptions that Whitehead made: his doctrine that discrete emergent "occasions" or elementary events[1] are the ultimate realities, and his view of "eternal objects" as potentials for the self-realization of these occasions.

Let us turn to the first assumption, namely, that event-like "occasions" or "actual entities" form the final real things which constitute the world.[2] This thesis implies an atomistic ontology. Whitehead himself says, "Thus the ultimate metaphysical truth is atomism," and he calls his philosophy "an atomic theory of actuality."[3] He takes the continuum of reality to be a phenomenon that is derived from the discretely emergent actual entities. Taken by itself, the continuum is only possibility, a "potentiality for division";[4] it is divided by the actual entities. With this claim, Whitehead enters into conflict not only with Isaac Newton's theory of absolute space and absolute time but also with Bergson. In Bergson's type of process philosophy, "duration," and with it a form of continuum, is fundamental. Bergson's "duration" is the continuum of becoming itself, while for Whitehead becoming is not continuous; in line with the paradoxes of Zeno, he asserts, "there can be no continuity of becoming."[5]

On this question, Samuel Alexander sides with Bergson—although Alexander, anticipating Whitehead's somewhat later remarks,[6] criticizes Bergson for setting up an opposition between space and time. The "spatialization" of time—which Bergson had judged to be the work of the understanding, and which he blamed in large part for the errors of traditional substance metaphysics—Alexander takes to be the essence of time itself.[7] Only space can make continuity possible, not only because a moment of time can be common to different places, but above all because, inversely, many consecutive events can occur at the same place.[8] A succession of instants by themselves would lack continuity: "[I]t would consist of perishing instants."[9] Continuity, which Whitehead later secured by his subtle theory of "eternal objects" and their ingression into the world of actual occasions, is guaranteed in Alexander's thought by a space which has not yet been relativized. To this extent we can understand Alexander as Whitehead's precursor.

However, Alexander's conception of the infinity of space-time, as the condition for the determination of finitude, is opposed to Whitehead's. Following Samuel Clarke and also Immanuel Kant's Transcendental Aesthetic, Alexander thinks of finitude and, above all, individual instants or points of time as limitations of infinity. "The infinite is not what is not finite, but the finite is what is not infinite."[10]

Like Emmanuel Lévinas today, Alexander refers back to René Descartes when he defines the constitutive meaning of the infinite as making it possible to conceive of the finite as such. This means, of course, that atomism cannot be advocated as the final metaphysical truth. Alexander shares Bergson's conception that movement is primordial, and hence that it is always holistic and continuous.[11] "Motion is not a succession of point-instants, but rather a point-instant is the limiting case of a motion."[12] Distinguishing between point-instants is the product of intellectual abstraction. "They are in fact ... inseparable from the universe of motion; they are elements in a continuum."[13] Alexander consequently moves close to Benedictus de Spinoza: Spacetime is the whole of that which exists, the infinite, which precedes all finite actuality.[14]

With Bergson and Alexander on one side, and Whitehead's event-atomism on the other, we are faced with two fundamental and alternative approaches to process thinking. To be sure, Alexander attempted to distinguish infinite space-time both from the category of substance and from that of a whole that precedes its parts,[15] the latter because, in his view, such a whole is always to be thought of as composed of parts. However, with regard to motion at least, he did speak of the whole of motion as being prior to its individual space-time instants.[16] In contrast, Whitehead proceeds from the ontological priority of discrete events or their components. But does he not thereby fall into the logical aporias of every atomistic metaphysics? Plato stated the case long ago: Without the One, the others can be neither one nor many; there would be absolutely nothing.[17] Many ones are many of the same (abstract One), but they are also many in relationship and so parts of a whole; if they do not form a totality, then they cannot be thought of as exemplifying the same One. In any case, an encompassing unity must be presupposed if atoms are to be conceivable as unities at all.

If I am not mistaken, Whitehead nowhere discusses the logical difficulties inherent in the systematic concept of atomism, although in *Adventures of Ideas* he does discuss various forms of atomism.[18] Whitehead distances himself from the sort of atomism—going back to Democritus, Epicurus, and Lucretius—from which the positivist interpretation of modern natural science is derived. Under such an interpretation, the atoms are only externally related, according to the principle of randomness; the relations are no less external in Newton's mechanics, where they express laws imposed from without by the will of God.[19] In contrast, Whitehead sees Plato as the originator of a way of looking at things which understands laws (and thus the relationships they regulate between things) as immanent in the things themselves.[20] Whitehead himself is inclined to this conception.

Even earlier, in *Science and the Modern World*, Whitehead attacks any description of relations between events that is given solely in external terms, which he finds in the usual accounts of space-time relations.[21] Instead, insofar as the individual events are constituted by the relationships in which they stand, these are internal to the events: "[T]he relationships of an event are internal, so far as ... they are constitutive of what the event is in itself."[22] According to Whitehead, the acceptance of such inner relationships requires viewing the individual event as subjective—a subjectivity which integrates the manifold relations constituting the event.[23] When this occurs, these internal relations represent themselves as acts of the event itself and, as such, are called "prehensions." In Whitehead's main work, *Process and Reality*, this concept stands at the center of his analyses, while the discussion of external and internal relations fades into the background (though even here "prehensions" are still defined as "concrete facts of relatedness").[24] Each individual event prehends all the other events of the world that it knows. The other events it encounters must be appropriated as its own: "[E]ach actual entity includes the universe, by reason of its determinate attitude towards every element in the universe."[25]

It may appear that the one-sidedness of atomism is thereby counterbalanced. The thesis that every individual event is conditioned by the totality of all the others does do apparent justice to the constitutive meaning of the whole for the individual. Still, one should observe that Whitehead does not speak directly of the universe having a meaning for the individual, but only hints at it indirectly, based on the relationship of every event to every other "element" of the universe. Only this latter sense is expressed when he writes, "[E]very actual entity springs from that universe which there is for it."[26] Since the universe (or the space-time continuum) is not given as a real whole to the individual event, it is always the individual event itself which must integrate into a whole the manifold relationships into which it enters. Consequently, as many perspectives on the universe arise as there are events that emerge.

It is no accident that Gottfried Wilhelm Leibniz comes to mind at this point. Whitehead explicitly refers to Leibniz's doctrine of monads: "I am using the same notion, only I am toning down his monads into the unified events in space and time."[27] With his thesis of the "window-lessness" of monads, however, Leibniz would have denied the concrete reality of internal relations. For Leibniz, the monads do not stand in real relationships to each other but only mirror the primary monad and the universe created by it, each refracting the universe from its own finite position. According to Leibniz, then, natural laws are as much externally imposed on the world (namely, by God) as they were in Descartes.[28] Whitehead, on the contrary, wants to understand the laws of nature as emerging out of the reciprocal relationships of the things themselves, as expressing these reciprocal relations. Hence individual events appear to him not only as reflections of the universe but also as the subjects of the creative integration of the manifold relations which constitute them. The spatiotemporal continuum is then taken to be the result of an abstraction from the concrete event itself, out of which the relationships among the "actual entities" emerge.[29]

In the end, Whitehead's theory of the prehensions does not really counterbalance the one-sidedness of atomism. For the whole of the universe (or of the spatiotemporal continuum), on Whitehead's account, has no inherent integrity of its own in the way that the monadlike events do. For Leibniz, the universe, established in the mind of God, precedes each individual creature and is only mirrored by the creature. In Whitehead's thought, however, God is not the creator but only the cocreator of all real events; consequently, the actual occasions, as self-constitutive, are themselves the ground of the continuum that expresses their "nexus"—the continuum is derived from them.

Whitehead appeals to the subjectivity of individual events as the central force that integrates the manifold of self-constituting relationships. Using this concept, he challenges the materialistic description of natural processes, which settles for merely external relationships, with a more profound vision.[30] But the idea of a self-constituting subjectivity of actual occasions leads to new difficulties. On the one hand, the actual occasion or actual entity ought to be the ultimate constituent of the physical universe. On the other hand, these ultimate constituents of the universe can be broken down further into the relations or "prehensions" which constitute them. Whitehead's response is that the analysis of an "actual entity" is feasible only in thought: "The actual entity is divisible; but it is in fact undivided."[31] If, however, the analysis of the actual occasion into the prehensions (or internal relations) which constitute it is feasible only through mental abstraction, then we are faced with a problem. How is it possible to continue to interpret the actual occasion itself as a process with different phases, in which it generates itself, while also asserting that the end phase of this process is identical with the complete duration of the event?[32]

The entire presentation of a "genetic division" in *Process and Reality*, with its differentiation of various phases in the self-constitution of the event,[33] raises the suspicion that Whitehead has confused the abstract and the concrete, thereby committing the "fallacy of misplaced concreteness," which he has often astutely criticized in other thinkers. Whitehead holds that we cannot, in fact, divide the actual occasion further but can only abstractly differentiate the relationships that constitute its identity. If so, he cannot in the same breath characterize the actual occasion as being the result of a process in which these relations, distinguishable only in the abstract, are actually integrated. Even more confusingly, the relations themselves are said to be constituted only by the actual occasion.

Yet such a position is clearly necessary if Whitehead wants to be able to affirm the subjectivity of actual occasions as *causae sui*.[34] If one conceives of the actual occasion as determined merely by the collision of the intersecting relations that constitute it at a moment of time, then the actual occasion is thought only as an object. As such, it cannot be separated from its field; it is nothing more than a singularity of the field ("the systematically adjusted set of modifications of the field").[35] It would seem that the subjectivity of the event must be assumed if the event's independence is to be at all conceivable. Thus, such subjectivity is a necessary condition of Whitehead's atomistic interpretation of reality. Now, if the independence of actual occasions is thought of as self-constitution, then it would seem to follow that the actual occasion itself must be reconceived as a process which integrates that which precedes it and thereby constitutes its own identity. This conception is self-contradictory, however, for the actual occasions are claimed to be the ultimate components of reality and not themselves integrations of more primitive components. This fundamental thesis cannot be reconciled with the assumption of the actual occasions' self-constitution.

Whitehead's "genetic analysis" of the actual occasion clearly amounts to an extrapolation which forces the experiential structures of more highly organized forms of life onto the interpretation of actual occasions. Whitehead himself described this procedure of his speculative philosophy as the method of imaginative generalization; and it is precisely the principle of self-constitution as "creativity" which forms, in his own philosophy, the central application of this method.[36]

Whitehead's doctrine of the subjectivity of actual occasions shares many individual features with the philosophical psychology of William James. More specifically, such features include, on the one hand, the momentary character of the ego and, on the other, the description of each ego-instant as being a momentary integration of experience, and especially an integration of the past of such ego-instants. Presumably, Whitehead's theory of the subjectivity of actual occasions can be interpreted, to a very large extent, as a generalization of the idea of the ego in William James' psychology, a generalization achieved by applying this idea to the interpretation of the foundations of physics. It was not for nothing that Whitehead placed James, alongside Bergson and John Dewey, among the thinkers to whom his chief work, *Process and Reality*, is especially indebted. In *Science and the Modern World*, he even compared James to Descartes as a founder of a new era of philosophy.[37]

The demonstration of connections such as these is certainly not enough to substantiate any objection to Whitehead's claims. The procedure of imaginative generalization obviously plays a considerable role in any formation of philosophical concepts. Whitehead himself says, however, that such a procedure has the character of "tentative formulations" and that it requires, along with inner consistency and coherence, confrontation with facts: "Speculative boldness must be balanced by complete humility before logic, and before fact."[38]

Measured by this yardstick, the use of the structures of subjectivity for interpreting actual occasions seems illegitimate. James' psychology of subjectivity has to do with a real succession of moments of experience, while Whitehead's thought cannot claim any such real succession in the genesis of the individual actual occasion. James' psychology of the ego can conceive each individual moment of experience as a new integration of previous experience because the successive moments are really distinct and because the relation of the later to the earlier (as the integration of human experiential moments) constitutes the special quality of human, subjective behavior in the medium of experience, reflection, and memory. For his part, Whitehead can appeal only to the factual universal relatedness of all events as the basis for applying the Jamesian model of subjectivity to the relationship that obtains between newly emergent events and all other events (including their predecessors).

It is doubtful that the similarities here have sufficient reach. According to James, the ego, which always emerges momentarily, in no way relates itself to all preceding events but only to those earlier experiences made present to it by memory. The human faculty of memory, however, is a highly specialized function which cannot, without further ado, be attributed to all natural processes. Moreover, the integration performed by the momentarily emerging ego is, in James, conditioned by the fact that the human body, on the one hand, and the "social self" (composed of the expectations of a social identity that face the individual), on the other, do not emerge in a pointlike manner. Instead, they represent continua, in relation to which the point-by-point synthesis of the ego can function as the principle of novelty and creativity.

Whitehead's speculative extrapolation of the principle of subjective integration that emerges moment by moment may overestimate the measure of uniformity encountered in the real world. The generalization of the structure of the human ego, as understood by James (but not tied to the problematic of the self as distinguished from the ego), paradoxically reduces the special status of the higher forms of natural evolution, especially the evolution of life, to the level of elementary processes. These more complex forms of natural evolution Whitehead merely describes as diversely ordered series, or "societies," of actual occasions. Because certain abstract structural elements are reproduced in each individual succession and systematically modified, the societies appear as stable unities without finally being such. In *Process and Reality*, the relatively brief treatment of this topic suggests that the ontological dignity of stable forms is considered secondary to the structure of actual occasions. Were we to suppose, however, that the correct description of actual occasions would, in principle, decipher the code for the formation of all higher forms (since such occasions are the basis for all higher forms), then we would fall into the mode of thinking of materialism, which is precisely what Whitehead wished to challenge.

The fact that the emergence of broader forms cannot be derived from the actual occasions which supposedly compose them shows once again that the unity of the field cannot be reduced to the elementary point-like events that appear in it. With respect to the metaphysical relevance of form as actuality, not merely as structure in the sense of Whitehead's "eternal objects," we also see how one-sided the atomistic interpretation of reality is: It cannot treat wholeness and individual discreteness as metaphysical principles of equal importance.

Remarkably, however, it is precisely Whitehead's genetic analysis of actual occasions with its paradoxes that offers constructive new perspectives upon which to build. According to Whitehead, the phases of formation are not to be thought of as temporally successive, since the actual occasion is what it is as an undivided unity. Therefore, representing the occasion as a process of formation appears paradoxical to us. But Whitehead's analyses do illuminate our understanding of processes whose phases certainly must be thought of as temporally successive, yet in which the goal of becoming [*das Werdeziel*] for the form has always been present. All life processes, for example, seem to be of this nature. In the process of its formation, the plant or animal is always this plant or this animal, although its specific nature comes fully to light only in the result of its formation. By way of anticipation, it is always that which it will become only in the process of its formation. The identity of its being is assuredly not that of a momentary occasion, but resides in the identity of its nature, of its essential form, which endures over the course of time. By anticipating its essential form in the process of its own formation, a being's substantial identity is linked together with the notion of process.

In Whitehead's genetic analysis of elementary processes, the concepts of "subjective aim" and "superject" play a similar role. Already in *Process and Reality*, Whitehead speaks occasionally of anticipatory feelings using the notion of subjective aim; he then expands greatly on this in *Adventures of Ideas*.[39] Even in the latter book he does not go so far as to describe the significance of anticipation for the formation of the subject, in the sense that its subjectivity is constituted out of its future, a future that already determines the present by way of anticipation. Rather, for Whitehead anticipation means that the subject, constituting itself in the present, includes also its future relevance for others (its "objective immortality") in the act of its self-constitution.

Whitehead did not thereby exhaust the theoretical potential of the notion of anticipation that is implied in the concept of "subjective aim." Aristotle's analysis of motion, which forms the background to all teleological descriptions of process, made fuller use of this notion. In the case of natural motions, Aristotle interpreted the anticipation of the final state within the moved object itself as entelechy. By doing so, he reinterpreted the effect of the future goal upon the present becoming along the lines of the influence that a living organism's seed has on its future goal. Aristotle nevertheless spoke of an effect of the end upon the process of becoming. Whitehead never speaks in this way because he sees becoming in each of its stages as self-constitutive. This is why, despite his use of teleological language, the element of anticipation cannot really become constitutive in his interpretation of subjectivity.

The idea of the radical self-creation of each actual occasion is the reason why Whitehead's metaphysics cannot be reconciled with the biblical idea of creation or (therefore) with the biblical idea of God. To be sure, American process theology has attempted to interpret Whitehead's concept of creativity in terms of a divine creative activity.[40] But in Whitehead's thought itself, the constitution of each actual entity's subjectivity remains always a self-constitution, and this despite the dependence of each actual entity upon God, who provides it with the conditions of its self-realization through its "initial aim." This stems from the fact that Whitehead ties the teleological structure of formation to actual occasions. He claims for them the character of processes but does not allow for temporal extension in the sense of a succession of phases in time: Actual occasions, which supposedly compose all else, are considered momentary and undivided.

The matter would be otherwise if we limited Whitehead's genetic analysis to processes that take place in time, instead of using it to explain the constitution of actual occasions. Then the "subjective aim" of the process would have to do with the actual, still-to-come future of each one's own essential completion. This future completion would not depend on present decisions alone, though it would eventually be reached or not reached by such decisions. Correspondingly, the anticipation of one's own essential completion in the future would gain greater significance for the constitution of subjectivity; the latter could no longer be identified with the self-creation of present decisions, but would be dependent on the whole of one's own essential completion being manifested in each present.

Such a conception would certainly no longer be that of an atomistic metaphysic. It would no longer attribute subjectivity to the randomness of actual occasions. Rather, from the impossibility of such attribution (because it involves the paradoxical assumption of a nontemporal process), we can construct an argument for the claim that the independence of finite being and subjectivity can increase along with the complexity of forms rather than be fully expressed at the outset in the elementary occasions. The unity of the field, from which actual occasions proceed, would no longer be traceable back to a network of relations that is itself first constituted by these occasions. Rather, we would have to conceive the unity of the field together with the unity (also underivable from but composed of actual occasions) of the forms that appear in increasing differentiation on higher levels of natural process.

Again, such a view of the matter would no longer be atomistic because it does not limit reality (in the sense of what is actual) to the undivided actual occasions. This is hardly reason enough, however, to remove such a view from the circle of process philosophies, even though it holds to the idea of an essential identity of that which continues to become throughout the process of its formation. The unity encompasses the whole process and so links the fundamental intention of the concept of substance with the process perspective. It is precisely in this direction that Whitehead's analysis of genetic processes, with his concept of the subject as the "superject" of its own process of formation, has provided important new impulses, even if these impulses bear fruit only after they have been liberated from the confinement of momentary actual occasions and the atomism which accompanies them in Whitehead's thought. In rethinking the matter in this way, we also free process thought from the aporias which have arisen within a theoretical context burdened by these assumptions.

# ⇒15⇐

# A Liberal *Logos* Christology

## *The Christology of John Cobb*

THE CHRISTOLOGICAL SKETCHES of the preceding decade usually begin with the historical Jesus, in order then to justify the Christological dogma as an exposition of the human-historical reality of Jesus but also critically to evaluate that dogma through the use of this criterion. The classical *Logos* Christology of the early church is not easily reconciled with such an approach, since conversely it interprets Jesus from God hither and hardly raises the question of how one is to justify the assumptions about the divine *Logos* from which it proceeds. The personalistic interpretation of the *Logos* Christology by dialectical theology, which found in the *Logos* only the hypostasized divine Word of revelation, has contributed rather to the discrediting of present attempts in this direction. For it becomes visible therein that the classical function of the *Logos* Christology—the exhibition of the universal validity of Jesus through the connection of his historical form with the philosophically accessible principle of world order—is not readily attainable in present theological thinking. Since 1964, I myself have criticized as a hopeless venture the connection of the figure of Jesus with the question of the order of the cosmos, a question now dealt with through the knowledge of laws by the natural sciences,[1] and I pleaded for a replacement of the *Logos* concept by the notion of revelation, which in any case concerns Christology's "point of departure."[2] Even so, as a second step, Christ must certainly be thought of as mediator of creation, for only thereby is he genuinely conceived as one with God, as participating in the all-determining reality of God.[3] And only thereby is the universal validity of the figure of Jesus to be legitimately grounded. This substantially amounts to a demand for a renewed *Logos* Christology, which interprets the historical figure of Jesus in continuity with God's relation to the world in general.

John Cobb has now produced such a *Logos* Christology, entitled *Christ in a Pluralistic Age.* It is noteworthy that precisely a liberal theologian like Cobb has taken such a step, for liberal theology has ordinarily concerned itself with stripping away the divinization of Jesus in the early church's Christology (or what appeared to it as such) as a secondary coat of paint over his true, historical humanity. Even Cobb will permit no diminution of the full humanity of Jesus.[4] His sympathy for Piet Schoonenberg's defense of the human personality of Jesus is probably grounded primarily therein (compare *CPA* 271, 166–167). But he correctly considers it necessary, beyond that, to ask about Jesus' distinctiveness, specifically in regard to his relation with God—in regard to a "concrete particularity of the divine presence and action in him as there is in each of us" (*CPA* 103–131). Only such a distinctiveness of his relation to God can ground Jesus' authority for us. Whereas Cobb was content in an earlier essay to characterize this distinctiveness anthropologically (compare *CPA* 13), it is now developed in the form of a *Logos* Christology (*CPA* 135; compare 138–139). Cobb arrived at this posture through the natural philosophy of Alfred North Whitehead, which enabled him to expound the universal significance of the words and work of Jesus with a forcefulness that is unattainable through a merely anthropological interpretation.

With such a procedure, certainly, there is even already presupposed a definite concept of the *Logos*, whose presence in Jesus is asserted. How Cobb justifies in an authentically theological manner his concept of the *Logos* as "the power of creative transformation" (*CPA* 131) does not become very clear from his statements. That is a problem which he shares with the *Logos* Christology of the early church. The initial chapters of his book, which exhibit the principle of "creative transformation" in the histories of Christian art and Christian theology, cannot furnish any theological justification for that as the *Logos* or messianic principle which appeared in Jesus. They certainly show the significance of the principle of "creative transformation" for the self-understanding of modern art, and also its actual efficacy in the history of modern theology. But derivation from the history of Christianity does not yet assure, here as elsewhere, that one thereby has to do with a Christian motif, much less the central Christian motif: Considering the ambiguity of all processes of secularization, it could also be understood as a falling away or detachment from its Christian origin. Accordingly, Cobb emphasizes correctly: "This account of Christ in art and theology cannot stand alone. Christ is indissolubly bound up with Jesus" (*CPA* 62). Part Two of the book is therefore supposed to demonstrate that the principle of "creative transformation" exhibited in Part One has in fact appeared also in Jesus Christ, and, to be sure, in him completely for the first time. But even here the basis for the applicability to Jesus of the previously developed concept of the *Logos* as "creative transformation" exists only in the reference to Jesus' consciousness of full power of authority as the authority of a man who dares to act in God's place. "This suggests that in some special way the divine *Logos* was present with and in him" (*CPA* 138). This grounding, on the contrary, already presupposes the previously mentioned *Logos* concept.

Now it could be easily asserted that Cobb's actual justification for the Logos concept which he employs is given not at all through historical-exegetical argumentation but on the basis of philosophy. Here the concept of the Logos is only another name for the cosmological function of Whitehead's God, namely, his "primordial nature," by virtue of which, according to Whitehead, God gives to every event its ideal possibility, its "initial aim" (*CPA* 225; compare 229, 76–77). Therein the idea of creative transformation is grounded, the new is brought forth, the future is granted (*CPA* 70–71), and thereby the possibilities of the past are realized (*CPA* 69)—therefore transformed without being destroyed. While this aspect of God, which for Cobb can be presupposed, is called Logos by him, he establishes only a conjunction with traditional Christian language which justifies theologically what would be philosophically required besides, namely, to describe structurally Jesus' fulfillment of existence in the sense of this general philosophical assumption. Does the Logos concept therefore furnish here only a justification for interpreting the figure of Jesus in the sense of Whitehead's metaphysical principles?

One will have to add at least that Cobb considers an inner, essential relationship between Whitehead's philosophy and the Christian faith to be a matter of fact (*CPA* 229). Cobb finds this essential relationship above all in the notion of divine love (*CPA* 229, 248). Of course, Whitehead in his principal work[5] first speaks of a divine love toward the world in reference to the divine preservation of the values independently realized by the finite processes, whereas Cobb already interprets as an expression of God's love the divine presentation of ideal future possibilities (the "initial aim") for the self-actualization of creatures. For this he refers not to Whitehead's principal work but to another of Whitehead's few references to the notion of God. In *Adventures of Ideas* the primordial nature of God is "pictured as the love that lures man to adventure."[6] Whitehead does in fact speak there of the divine love, though not in reference to God's relationship to the creation but in the sense of an erotic impulse in God himself toward the actualization of ideal possibilities, whereby God's infinity is actualized in the world process.[7] One must judge that this notion is not simply identical with the biblical notion of the forgiving love of God, for this is creatively bestowed upon the world, whereas Whitehead has rejected precisely the notion of creation.[8] His concept of divine love points on the one hand to God's own essential realization, but on the other hand, as love toward the world, it points to the preservation of the values produced by the finite processes themselves.

A similar idea results in regard to the reference to the future, which is central to Cobb's concept of "creative transformation" (*CPA* 70–71). The Christology relies here on the detailed explanations in *God and the World* (1969) concerning the divine "call forward" toward a possible fulfillment of creatures in the future (*GW* 42ff.). God is that one (*GW* 63) who "calls us forward in each moment into a yet unsettled future" (*GW* 55). Cobb connects this assertion with the notion of God's love: God is "a lover of the world who calls it ever beyond what it has attained by affirming life, novelty, consciousness, and freedom again and again" (*GW* 65). This notion is undoubtedly impressive, and it seems to me to be thoroughly in agreement with the understanding of God in the Bible. But can it really be attributed to Whitehead, as Cobb believes (*GW* 66)? I doubt it. So far as I can see, Whitehead has given the future no ontological status, not even in his doctrine of God. He has even explicitly said of love that it does not concern itself about the future.[9] Therefore the correlation between love and hope has simply not been thematized by Whitehead. And although in *Adventures of Ideas* he also depicts the creative synthesis of all events as an "anticipation," and is even conscious of the new qualification of the present through the future that is grounded in that synthesis,[10] the thesis of a constitutive significance of the future for what is present, as suggested by the term "anticipation," does not lie within Whitehead's field of vision. One may not consider unimportant the fact that certainly this aspect is *implicitly* present in Whitehead's formulations. It is not accidental that Whitehead has not developed these implications. Had he done so, he would have had to change his concept of actual entities as present or past occurrences. But that would have required a revision of his entire system. On the other hand, precisely because Whitehead has not investigated the constitutive significance of the future for the present, he has developed his problematic doctrine of timelessly subsisting abstract possibilities (eternal objects) over against what is actually real (actual entities). The timelessness of eternal objects actually *replaces* the missing reflection upon the constitutive significance of the future for present and past events.

How then does Cobb, in his striving for a theology along a Whiteheadian track, come to his vision of the God who calls the world to its future and *therein* is lovingly devoted to it? It appears that he has read Whitehead through the spectacles of the so-called theology of hope and of Pierre Teilhard de Chardin.[11] Thereby Cobb could rightly suppose that he is only working out explicitly the specific matters which are somehow implicitly present in Whitehead. But so far he seems to be overlooking the fact that such an explication amounts (at least implicitly) to a revision of the metaphysical foundations of Whitehead's thinking. Without doubt one can also thereby learn yet a great deal from Whitehead, but in such a way that one moves beyond Whiteheadian scholasticism, which appears to me still to be a danger in process theology.

This is not the place to develop more exactly the suggested revision of the foundations of Whitehead's philosophy.[12] It is sufficient in the space of this contribution to point to the tensions arising in Cobb's Christology from the fact that, on the one hand, he develops a theological vision stimulated by Whitehead but leading beyond him, while, on the other hand, he narrowly attaches himself again and again to theorems expressly expounded by Whitehead.

The first difficulty of this sort already steps forth in the introduction of the concept of the *Logos* as "creative transformation" without justification from the historical particulars of Jesus' own history. Cobb's expressed vision of "creative transformation" as origin of the contingently new in the world, but also as an expression of God's love toward the world—which works in Teilhard's sense as a creative unification by overcoming the isolation of the individual—would be thoroughly appropriate for such a justification.[13] It could be developed out of the biblical understanding of the historicity of God's relation to the world, and out of the revelatory function of the history of Jesus which is to be decided in connection with the whole of this divine history. But Whitehead's philosophy suggests another conception of "creative transformation," which conforms to Whitehead's explicit statements about the initial aim presented by God to each event for its self-constitution. Now this initial aim, taken for itself, is not at all futural but is an ideal possibility, an "eternal object." It first becomes the transformation of extant reality through the way in which the self-constituting occurrences, through their own creativity (!), include their initial aim into the process of their self-constitution, making it their subjective aim. And this concept certainly is not grounded in the history of Jesus. It can only be substituted for it in the sense of a positive analogy.

Through this substitution, Cobb's Christology gains significant advantages. Whitehead's philosophy is one of the few detailed philosophical interpretations of the natural world which our age has produced, and among the sketches put forward it is perhaps the most significant. Of course, one may not thereby close one's eyes to the one-sided partialities, clinging even to it, as has so often happened in the history of theology where one has made a decision for this or that philosophy and then followed its theorems dogmatically. Cobb rightly considers it to be a goal of his book "to seek a positive account of the kind of experience that is expressed in varied religious contexts rather than to reject them as incomprehensible or as doing violence to the facts" (*CPA* 117). The question is only whether his close attachment to Whitehead's philosophy always allows him to pursue this objective. In that he connects the figure of Jesus with Whitehead's philosophy through the concept of the *Logos*, the universal cosmic significance of the history of Jesus becomes articulatable in an imposing way. That must be a goal of all Christology. But perhaps Cobb, through the wholesale acceptance of Whitehead's natural philosophy, attains this goal too easily, without the difficult work of a critical transformation, in the light of the history of Jesus, of the foundations of the philosophical system from which he proceeds. The ease with which Cobb attains the cosmic perspective of a *Logos* Christology could be purchased at too high a price, namely, on the one hand, through the fact that particular tensions between Cobb's authentically Christian statements and Whitehead's fundamental philosophical ideas are only verbally disguised (as has been shown above in respect to the notion of love and the efficacious working of God as originator of the new), and, on the other hand, through the fact that particular aspects of the history of Jesus must be eliminated, especially the realism of his proclamation of the near end of this aeon with the arrival of the kingdom of God. Indeed, John Cobb seems to perceive just therein a superiority in Whitehead's thinking, in that his philosophy permits theology to be released from problems of the end of the world that are bound up with Jesus' imminent expectation, since this imminent expectation has apparently been refuted by the continuation of history (compare *CPA* 225, 250). But not only for contemporary Christendom does clinging to the notion of an end of the world and of history stand in tension with the non-Christian understanding of the world. On the contrary, that was already the situation of Christian thinking vis-à-vis the classical systems of Greek philosophy, and precisely the adherence to a realistic eschatology, and therewith to the notion of an unrepeatable historical process irreversibly running toward its end, enabled Christian thinking to achieve a "creative transformation" of the philosophical heritage of antiquity.

The most important limitation in Cobb's diagnosis of the history of Jesus through the philosophical scheme of Whitehead might consist in the fact that the kingdom of God is no longer presented in the dynamic of its coming, as the field of force of the divine future determining the present.[14] Cobb himself judges this dynamic to be broken for us, two thousand years later: "We do not anticipate an imminent coming of a new order or a new age in which the ambiguities of our world will be superseded" (CPA 225; compare 249ff.). It is true that the imminent expectation of Jesus and primitive Christianity has not been fulfilled in its original sense. But in this connection Cobb curiously does not consider the event that made this delay of the eschaton bearable for early Christianity by guaranteeing to believers the presence of future salvation already now, in the midst of this world of death: the Resurrection of Jesus. One can say that the imminent expectation of Jesus has thereby encountered a transformation, grounded in historical experience, which nevertheless remained in fundamental continuity with Jesus' message of the world-altering dynamic of God's future. Instead of following this development and interpreting its meaning, Cobb tries to formulate the presently authoritative content of Jesus' message through immediate adherence to Whitehead's conceptual standards, especially to his concept of the kingdom of Heaven. Although one can certainly agree with Cobb that this notion of Whitehead has been deeply influenced by the New Testament, I am not able to acknowledge that it is "homologous" (CPA 227) with the New Testament. Cobb sees the antithesis himself: "In the New Testament the Kingdom of Heaven will come no matter what we do. For Whitehead it will preserve and redeem our actions no matter what they are" (CPA 227). In the New Testament the futural kingdom of God is the majesty of the coming God himself, from which nothing that is present can escape. It is not only a guarantee for the immortality of the outcome of all temporal events in spite of their transitoriness. Cobb tries vainly to reconcile this antithesis by indicating that in both conceptions the human decision has a central significance (*CPA* 227).[15] That is the case for many views of the world, but what matters is the character that is ascribed to this decision before which humanity presently is placed. In the New Testament, this is the majesty of the God of Israel, to whom alone the future belongs, so that only through being joined with him can humanity hope for salvation. On the other hand, Whitehead's God preserves, in one way or another, the positive content of our experiences and actions. Accordingly, there is no interest in God's coming, in his future.

Only at one point is Jesus' orientation to the future of God's Lordship in accord with Whitehead's philosophy, and Cobb propels this point again and again into the center of his interpretation: that is, the implicit futurity of the ideal possibility for realization that is presented, according to Whitehead, to every occurrence. The person of Jesus is accordingly represented as an exception to the general circumstance that every occurrence makes the ideal possibility presented to it (initial aim) into the object of its own self-actualization (subjective aim). Cobb finds the uniqueness of Jesus in the fact that Jesus has made his God-given initial aim, the *Logos*, into his subjective aim without distortion (*CPA* 140ff.). Therefore in Jesus is not found the "usual tension between the human aim and the ideal possibility of self-actualization that is the Logos" (*CPA* 139–140). In this sense Jesus' "I" is constituted—or rather co-constituted[16]— through the presence of the *Logos* in him (*CPA* 141), and this means for Cobb that in Jesus there was no tension, no opposition of his "I" to God (*CPA* 140–142).

Part Two of Cobb's Christology ("Christ as Jesus") reaches its zenith in these explanations. Cobb began here with Jesus' words (chap. 5), in order thence to press forward through the question of their efficacy (chap. 6) and of the basis for their authority (*CPA* 131ff.) to the distinctiveness of Jesus' person (chap. 8). However, in this process of argumentation—for which the parable of the pharisee and the publican serves as a guide—Jesus' message of the kingdom of God is now limited to the question of Jesus' personal relationship to God. Thereby the future of the kingdom of God for humanity shrinks to the futurity of the God-given ideal of individual self-actualization for

Jesus' person. To be sure, Cobb also says something about the content of this ideal of existence, which became constitutive for Jesus' self-consciousness. In conjunction with Milan Machoveč he states that Jesus not only proclaimed the kingdom of God but "*incarnated this lived future* in his entire being" (*CPA* 138). However, Cobb does not engage in a more exact analysis of this fact, but employs it merely as an illustration for the formal thesis, then developed in orientation to Whitehead's doctrine of the subjective aim: "Jesus existed in full unity with God's present purposes for him" (*CPA* 141). Thereby Jesus' unity with the kingdom of God that he proclaimed is simply characterized as unity with the *Logos*, in the sense that "his very selfhood was constituted by the Logos" (*CPA* 139). However, a more exact analysis of the matter of the relation of Jesus to his proclamation, described in connection with Machoveč, would have to demonstrate, *first,* that the mission of proclaiming God's kingdom which constituted Jesus' self-hood had a content *distinct* from Jesus' individual existence, namely, the future of God's kingdom as it concerned the whole world; and *second,* that it thereby concerned not simply the *Logos* but the *Lordship of the Father.* But Jesus knew himself to be in no way identical with the Father. It is erroneous to conceive of Jesus' relation to God as such an identity that in it there "would not be the confrontation of an 'I' by a 'Thou'" (*CPA* 140). In his humanness Jesus knew himself to be so very different from the Father that he rejected the address "Good Teacher" with the remark: "No one is good but God alone" (Mark 10:17–18). This is central for the whole of Christology; for according to the Gospels, that Jesus had made himself equal to God was precisely the accusation of Jesus' opponents. His self-distinction from God, the Father of the coming kingdom, was the condition for Jesus' unity with God, which he had as "the Son" in obedience to the Father, and had in only that way.[17]

Unfortunately, Cobb has not taken notice of this matter which is so fundamental for the whole of Christology. He has not gone into the fact that Jesus' relation to God is first of all a relation to the *Father* and, as such, should be characterized by means of differentiation. Only in this roundabout way does Jesus' own unity with God as "Son" become understandable without diminution of his full humanity. Insofar as Cobb endeavors to think of Jesus' relation to God as an immediate relation to the *Logos*, along the track of the Christological debates of the early church, he not only misses the historical phenomenon of Jesus' existence but also gets entangled in the dilemma which has accompanied the Logos Christology of the early church since the conflicts of the fifth century, the dilemma of Monophysitism and Nestorianism. It is a self-deception if Cobb intends through process philosophy to be fundamentally beyond this dilemma that penetrates the entire Christological tradition. This dilemma has not only been a consequence of "substantialist" categories of thinking, as Cobb presumes (*CPA* 167). It is rather the consequence of every attempt to think of Jesus' unity with God immediately as Jesus' relation to the *Logos*, instead of in reference to Jesus' relation to the Father. Cobb is closely connected thereby with the model of the ("Nestorian") Christology of separation. It is significant in that regard that the *Logos* can only have value as "co-constitutive" (*CPA* 141) for Jesus' selfhood, because Jesus "freely chose to constitute his own selfhood as one with this presence of God within him" (*CPA* 173). Which is the subject of such a free choice? The man Jesus already united with the *Logos*? Then his unity with the *Logos* is not first grounded through this act of choosing. Or was the man Jesus, not yet united with the *Logos*, the subject of that free choice? Then the man Jesus was not one with the *Logos* from the outset. Christology avoids this dilemma only when the *Sonship* of Jesus is understood as a qualification of his whole person in the light of his relationship to the *Father.*

Jesus' self-distinction from the Father is also central to the Christian doctrine of the Trinity, and here also the inattention to this fact, which is to be exegetically established, has problematic consequences in Cobb's conception. To be sure, his theology has acquired a fundamentally positive relation to the doctrine of the Trinity through the introduction of the *Logos* concept (compare *CPA* 13). As Jesus is seen together with the *Logos*, so the Spirit is seen together with the future of the kingdom of God, so that the Spirit is the presence of this coming kingdom of God as Christ is the presence of the *Logos* (*CPA* 261–262). This idea is convincing, but it does not by itself establish a complete doctrine of the Spirit—as Cobb himself says (*CPA* 263). The distinc-

tion of *Logos* and Spirit as thus formulated is suggested by Whitehead's distinction of the primordial and consequent nature of God.
On the other hand, no point of departure is produced by Whitehead's
doctrine of God for the distinction of the Son (as also of the Spirit)
from the Father, nor for the biblically attested relationships between
the three persons. Therefore it is probably not accidental that these
elements are also missing in Cobb. It would be difficult to arrive at a
fully developed doctrine of the Trinity without a thorough revision of
Whitehead's metaphysics. Especially serious is the lack of a distinction
between the union of Father and Son appropriate to the historical facts
concerning the appearance of Jesus. This lack manifests itself in the
sentence already quoted, that for Jesus' mode of existence, united with
the *Logos*, no personal oppositeness to God, as an "I" to a "Thou," is
to be assumed (*CPA* 140); and the same lack appears in Cobb's critical reflection upon the development of the doctrine of the Trinity, that
perhaps only the Father should have been conceived as a hypostasis,
with the Son and the Spirit on the contrary as two modes of his activity (*CPA* 260): Such touches of a Dynamistic Monarchianism might
be inferred from Whitehead's statements about God. A Christian trinitarian doctrine of God would no longer be attainable along this path.
Not only the point of departure of the trinitarian understanding of
God but also Jesus' own divine Sonship depends on the personal self-
differentiation of Jesus from the father.[18]

The interpretation of the figure of Jesus as incarnation of the *Logos*,
in the sense of Whitehead's doctrine of the initial aim, is perhaps also
responsible for the restriction of Cobb's Christology to the person of
Jesus as that is expressed *in his words,* in distinction to *the history* of
Jesus. The relation between the appearance of Jesus with his message
on the one hand and his fate on the cross (as also his Resurrection) on
the other hand is curiously not discussed in this Christology, although
in 1969, in *God and the World,* Cobb had found in Whitehead an
access to the theological understanding of the cross as an expression of God's participation in the suffering of the world (*GW* 97). It
is unfortunate that this notion, generally held in Christology, has not
been developed further in reference to the suffering of the Son sent by
God to proclaim the nearness of his kingdom, and in reference to the

significance of this suffering for the world. To do that, it would certainly be necessary at least to work out the connection between Jesus' "words" and his specific authority more precisely than has been done by Cobb. To be sure, Cobb emphasizes incidentally, in conjunction with his exegetical authorities (*CPA* 104), that Jesus' notion of love is grounded in his eschatological message of the coming of God's kingdom. But he does not enter into the question of the extent to which the same holds true for Jesus' claim of authority.[19] Unfortunately, this question is often overlooked even in the exegetical literature, as Jesus' consciousness of the full weight of authority is dealt with as a remnant of the old dogmatic view of an immediate divine presence in him. But on this question hangs the possibility of recognizing the connection that leads from Jesus' message to his crucifixion: his claim to authority was inextricably bound up with the ambiguity that it could seem that he made himself God. But this ambiguity, arising in consequence of the appearance of Jesus, is not analyzed by Cobb, nor is the confirmation by the resurrection of Jesus' claim to authority. The entire historical process which is bound up with Jesus' claim to authority is displaced by the already discussed expositions of Jesus' structure of existence.

The history of the efficacy proceeding from Jesus—in which Cobb returns to the question of Jesus' authority—also becomes thematic only in a singularly reduced form, namely, in a reduction to the problematic of the salvation of the individual. It is concentrated upon justification, which is first presented in Jesus' parable of the pharisee and the publican (*CPA* 170ff.), then constitutes the key to the description of the efficacy proceeding from Jesus (chap. 6, "Life in Christ"), and finally is taken up once again in reference to the explanation concerning Jesus' authority (*CPA* 143–144): These now appear as an attempt to ground precisely the same efficacy of Jesus which emanates from that parable (*CPA* 143). In particular, Cobb develops a series of noteworthy insights on the theme of justification—namely, on the dialectic of publican and pharisee (*CPA* 109), the dissimilarity between the Pauline notion of justification and that of the parable (*CPA* 111), and the present ineffectiveness of justification faith (*CPA* 113–114). Cobb tries to provide an answer to this latter experience through a consideration of the Pauline notion of a "field of force" (*CPA* 116ff.) proceeding from Jesus Christ, which occasions in us a conformation to Jesus and so also a participation in his righteousness (*CPA* 122–123). However, thoughtful and beautiful though these explanations may be, it yet remains peculiar that Cobb limits the question of the efficacy proceeding from Jesus completely to the individual's self-understanding and scarcely goes into the rise of the *church* as a consequence of the history of Jesus. To be sure, it is occasionally mentioned that in early Christendom the church—alongside of the teaching of Jesus and bound up with it—was experienced as an expression of the "field of force" arising from Jesus (*CPA* 128ff.). But this historical observation remains inconsequential. Subsequently it is once again mentioned that the kingdom of God proclaimed by Jesus was connected with the rise of a new human community (*CPA* 222), but again no connection is established with the factual rise of the Christian church and its history. Elsewhere Cobb even emphasizes the systematic circumstance that faith and hope *require* a community "that expresses the field of force generated by Jesus" (*CPA* 185). But again there is no talk of the church in this connection. Instead of this, in connection with Paolo Soleri, the twelfth chapter develops the utopia of a new city which is characterized by an intensive living together of human beings.

The neglect of the church as the decisive, world-historical effect of the history of Jesus—if one is already inquiring into the effects proceeding from Jesus—is all the more astonishing inasmuch as Cobb opposed with sharp criticism (reminiscent of Teilhard) the phenomenon of a Christian individualism. He says that the call of Jesus, which was directed toward a "community of perfect openness," has *de facto* brought forth instead "the strongest and most isolated individuals in history" (*CPA* 110), who have brought the whole planet into the orbit of that history which measures time from Christ's birth. According to Cobb, this individualization united with alienation (*CPA* 184–185) must be overcome (*CPA* 42)[20] by a turning from I to we (*CPA* 215; compare 220). But Cobb does not go into the fact that the religious independence of the individual, grounded upon the justification faith (*CPA* 109), is only a product of Protestantism and an expression of specifically Protestant piety. His presentation, oriented toward the parable of the Pharisee and the publican, of the "Christian structure of existence" (*CPA* 109) as a continually renewed effort to overcome one's own self-justification does not by any means simply describe Christian piety, but is characteristic only for a quite specific type of Protestant pietism, which certainly is also characterized by the individualism criticized by Cobb. But how does it happen that Cobb considers this Protestant pietism simply to be the Christian form of existence? Does not even his own presentation remain closely bound up thereby with the individualism that he criticizes? Does he perhaps come to terms more with his own tradition of piety here than with the objective, historical phenomenon of Christianity and its development? Cobb's criticism of the isolation of the individual, as it has actually evolved in this Protestant piety, would not at all need to call up the Buddhist dissolution of the self as the corrective. This criticism could be grounded within Christianity upon the continually effective presence of the Catholic tradition and its orientation in the life of the church. It is interesting that such a notion lies further than Buddhism from Cobb. He could have exhibited the unintended loss of the unity of the church in the Reformation as a presupposition for the development of the individualistic piety correctly criticized by him. He could have concretized his legitimate longing for a new form of Christian community into the demand for an ecumenical renewal of a church encompassing all Christians. But instead, Cobb finds a symbol of his hope for a new community in the utopian city planning of Paolo Soleri. However, must not an inner renewal of human social life precede every modification of its external form, if this is not to work dehumanizingly? And in the understanding of the Christian tradition, is not the Spirit, through which the kingdom of God already becomes present (*CPA* 261–262), given to the church? Should Cobb's legitimate critique of the individualism of a privatized Protestant piety not lead to the notion of penitence, in respect to the continuation of the separation of the church? In any case, this may be the only field where theology and Christian engagement today could yet—beyond bare rhetoric—create really fundamental changes which, in their consequences, could even change society. Perhaps the reason that Cobb did not thematize the church as an object of Christian hope is to be sought in the fact that for him the church is an expression of the particularity of Christianity, and therefore does not seem to be sufficient as a criterion of the universality that corresponds to the cosmic relevance of Christ as incarnation of the *Logos*. The application of this criterion permeates Cobb's entire book. Already the generalization of the title "Christ" in the first two chapters, which at first glance seems strange, exhibits this tendency: Even where Jesus Christ is no longer an object of artistic representation, he remains efficacious as the principle of "creative transformation" determining the artistic productivity.[21] And even where Christian theology relativizes its own Christian tradition and religion, with the desacralization of the tradition it remains true to the Christian spirit. In contrast to the closedness of its own cultural tradition, cultural pluralism thus becomes the expression for the universal validity of Christianity: precisely in its engaging in the pluralism of the present situation of humanity it confirms its universal validity. I wish to agree with this tendency in Cobb's exposition and I consider in this sense even his sketch of a *Logos* Christology to be a positive contribution in the right direction, in spite of all the individual criticisms expressed. The original function of the Logos concept in Christology was indeed to make explicit the universal relevance and truth of the confessions concerning Jesus Christ. In a similar way, Cobb correctly turns also against every attempt to shield the Christian faith against critical discussion through "special pleading" (*CPA* 26–27), to assure it of independent certainty (*CPA* 238–239). Perhaps even the ecclesiastical form of Christianity appears to him as an expression of such a particularity shut in upon itself. I do not believe that such a judgment would be correct. A Christianity without a church can only lead to the impasse of individualism or to the identification of Christianity with a conservative or revolutionary civil religion. The church itself must receive the element of pluralism into itself. That is, primarily, in regard to the plurality of the Christian confessional traditions, the ecumenical task of the present, perhaps the greatest task and opportunity of the present generation of Christendom. But an ecumenical church, which, mindful of its own provisionality in comparison with the eschatological future of God's truth, has taken into itself an element of pluralism, will also find a new relationship to the extra-Christian religions and cultures of mankind—as Cobb discusses again and again, especially in regard to

Buddhism—and, to be sure, without renouncing its own identity. Only in this way will the Christian church be able to be in the full sense a "symbol and instrument of the unity of mankind," as the Second Vatican Council and similarly the World Council of Churches in Uppsala (1968) have characterized its essence.

The becoming of a new form of a creedal church must be the central object of the inner-worldly hopes for the efficacy of Christ. Cobb rightly places the theme of hope at the middle of the third and concluding part of his Christology. Regarding this subject—without debating the facts of the matter more closely, to be sure—what is Christologically at stake thereby is the return of Christ and the present efficacy of the LORD who will come again and who has already been elevated to Messiahship in the hiddenness of God. The locus of such present workings of the Christ who comes with God's future is primarily the church. Therefore the Christian hope must be directed toward the fact that the Spirit of the coming Christ unites and renews the church. This is unrestrictedly bound up with the hope—which Cobb designates as transcendent—in the kingdom of God and the resurrection of the dead. It is surprising that Cobb considers as competitive (CPA 243–244) the different contents of hope, the immanent and transcendent forms of Christian hope as well as the hope in the kingdom of God and in the resurrection of the dead. Hope for Christ's bringing about the unification and salvation of his church—with which, to be sure, Cobb does not deal—is closely connected to hope for the kingdom of God and for the resurrection of the dead. But especially these last two contents of hope cannot be separated from each other without losing their full meaning. In the interconnectedness of the resurrection hope and the kingdom of God hope, the interdependence of the individual and social destiny of humanity finds its expression.[22] And just because the union of these two aspects is definitely realized in no political order, the church is required as a symbolic representation of that messianic future of humanity. A competition between these contents of Christian hope cannot exist except, at most, in relation to Whitehead's special interpretation of the "kingdom of Heaven" and to Soleri's utopia of a futural city, insofar as these take the place of the church as a prefiguration of the reign of God.

# →16←

# A Modern Cosmology
## God and the Resurrection of the Dead

FRANK TIPLER'S omega point theory of physical cosmology rests on three assumptions.[1] The first and most important is the *anthropic principle* in its sharpest form as *final anthropic principle,* according to which life and intelligent life are not only a necessary part of our universe but also, after their appearance, cannot disappear again, but are destined to penetrate and rule the entire universe.

The second assumption is that the expansion of the universe, the existence of which, according to the standard cosmological theory, began with a big bang about fifteen billion years ago, will not continue unhindered; but, due to the influence of gravity, the universe will move into a phase of contraction that will ultimately end in a big crunch, a collapse of the matter of the universe into the smallest possible space—analogous to the "black holes" that already happen within the universe through the collapse of matter. The expansion of the universe neither continues as "open" into a constantly widening space nor is it "flat," as most cosmologists currently assume, in a balance between expansion and gravity; instead, it is "closed," with a universal collapse at the end. That is the only reason that there is an endpoint to its history, the omega point.

Tipler's third assumption is that the energy available in the universe is limitless. Therefore, the end will not be the "death of heat," a condition of maximum entropy, but possibly eternal life, that is, maximum use of information. For, according to Tipler, life *is* the storage of information (*Physics* 163ff.). On the way to the omega point, life must penetrate the whole of the material universe and govern it. But the omega point itself is the place of maximum storage of information and thus the point in space-time that is both immanent and transcendent (*Physics* 199). This, says Tipler, gives us the characteristics of personality: omnipresence, omniscience, omnipotence, and eternity (*Physics* 198ff.).

These characteristics of the omega point make the ultimate future of the universe also the place of its creation. Thus, the time-perspective even of Tipler's description of the history of the universe is reversed: God, as the ultimate future of the universe, is its creator, who in the course of the universe's history draws his creatures into communion with himself. While we act out of our present toward the future, because we have the future outside ourselves, God, who is himself the absolute future, places his creatures in an existence that precedes that future and moves toward them.

Tipler rightly asserts that his statements about the omega point correspond to what the Bible says about God. The God of the Bible is not merely related to the future through his promises; he is himself the future salvation that forms the core of the promises: "I will be what I will be" (Exod. 3:14). He is the God of the coming reign, who, while already master, in a hidden way, of the world as his creation, will be fully revealed only in the perfected future, in his rule and, thus, in his divinity. That is the reason the future of the reign of God is the core of Jesus' message, and the object of his prayer: "Your kingdom come" (Luke 11:2).

Thus, Tipler joins together the fundamental statements of the traditional Christian doctrine of God: Omnipresence, omniscience, and omnipotence are closely linked to the idea of an ultimate future as the place of maximum information. And in doing so, Tipler rightly distances himself from a notion of God as a spirit modeled on our human spirit, for "a spirit similar to the human spirit is the manifestation of an extremely low level of processing of information" (*Physics* 200).

God's omniscience infinitely exceeds the forms of our knowledge and is to be thought of in intimate union with God's omnipresence. As the end of space-time, the omega point is both immanent and transcendent to every point in space-time (*Physics* 299; compare 36–37). The same was emphasized by classical theology in the notion of God's omnipresence. Likewise, the idea of God's omnipotence and eternity contains the unity of immanence and transcendence. Tipler rightly understands God's eternity not as timelessness in contrast to all time, that is, in the symbol of a one-sided aspect of transcendence, but with Boethius as the unlimited possession of all the things that are for us separate in time, in a single present (compare *Physics* 202–203).

Since the God of the omega point is a being of maximum information storage, the notion of God's personhood offers no difficulties for Tipler. Because he conceives of person in terms of the ability to communicate (*Physics* 46–47; compare 200), he looks back to the Greek concept of *prosopon* as "face" or "mask" and takes up the idea of a plurality of persons in God (*Physics* 201). Here is at least an openness to Christian teaching on the Trinity, despite his reserved remarks on this topic (*Physics* 379ff.) in connection with the person of the Son and its position in the Trinity.

Tipler's relationship to the doctrine of the Trinity is, of course, dependent on his stance regarding Christology. We will come back to that. In any case, there is a fairly broad agreement between Tipler's statements about the characteristics of the omega point and Christian doctrine of God.

Does this mean, as Tipler has sometimes claimed, that theology dissolves into physics? In light of his omega point theory, I would rather speak of an *approximation* of physics and theology. The theory, with its beginning in the anthropic principle and its assumptions about the future of the universe as a closed whole, as well as about the unlimited increase of storage of knowledge en route to the omega point, is something like an introduction to the idea of God as the absolute future of the universe. It is only with the omega point that the viewpoint reverses: The last becomes first, the endpoint is the creator of the universe. But for now that remains more theology than physics, even though Tipler doubtless succeeds in developing a connected argument that permits the joining of the idea of creation as well as the eschato-logical hope for the resurrection of the dead with the characteristics of the omega point as the ultimate future of the universe.

When Christian theology thinks of the universe as "creation," that is a description of the universe in terms of God, and not the reverse, describing God in terms of the universe. The fundamental statement here is that the existence of the universe—seen from God's point of view—is "contingent." That means that the existence and nature of our universe are, from God's standpoint, not necessary. They could be different, or they could simply not be. The idea of God contains the thought that Godself cannot not be: If God exists, God is necessary in and of Godself. But the universe is contingent. Its existence is an expression of God's free decision. To that extent, it is creation. If the universe existed "necessarily," as God does, it would be an eternal cor-relative of God, and the existence of the universe could then not be an expression of God's freedom and love as creator but would be a condi-tion of God's own identity, a condition not subject to God's power.

According to Christian teaching, of course, there is only one uni-verse, not a multiplicity of universes in the sense of one particular inter-pretation of quantum mechanics, Hugh Everett's 1957 many-worlds hypothesis (*Physics* 214ff.). In my view, but also in the judgment of many physicists, this many-worlds hypothesis involves a problem-atic ontologizing of the plurality of alternative states, which accord-ing to quantum theory may take up a given particle at any moment. Tipler writes in his book that he is persuaded of the plausibility of the many-worlds hypothesis by Friedrich von Hayek's writing on the con-cept of capital. According to Hayek, "the one correct definition of the wealth a society owns" is "a complete list of the alternative incomes it might achieve in the course of time with the means available" (quoted in *Physics* 219).

But that is about *possible* alternative investments, which may *not* all be equally realized. In the same way, quantum field theory may have to do with alternative possibilities that cannot *all* be realized simulta-neously. The plurality of alternative possibilities, then, does not ground any real multiplicity of worlds.

In Christian teaching, the uniqueness of the universe is connected to its origin in the creative love of God, which decided to create, out of all the worlds that might have been, this one universe. Moreover, in Christian teaching, the idea of the love of God as motive for creation links the eschatological fulfillment with the act of creation, for the resurrection of the dead, to which Christian hope is directed, expresses that the eternal God holds steadfastly to his creation, will not let it go, will not abandon it to death. Human beings in particular are destined for eternal communion with God; therefore, God will raise them from death and so transform them through judgment that they are able to participate in his light and his glory.

Especially in the matter of the raising of the dead, Tipler again comes very close to Christian doctrine. The limitless storage of information at the omega point allows, in combination with God's omnipotence, for the identical repetition of what has been, according to the model of a computer simulation. This does not involve a material continuity or identity with the previous bodiliness, but Christian doctrine does not demand that. Even in this life, the material components of our body are in constant flux. What is crucial, as Thomas Aquinas, following Origen, already emphasized, is the *form* of our bodily existence that is programmed in the soul.[2] At the same time, it should be recalled that communion with the eternal God requires a transformation of our earthly form of existence, as Paul says: This mortal thing must put on immortality (1 Cor. 15:53). This is already part of the idea of sharing in eternal life, and the transformation of our life in its participation in the eternal life of God implies the judgment, the purification of everything that cannot exist in the presence of the eternal God.

Incidentally, Tipler sees the motive for the eschatological raising of the dead in God's selfless love (*Physics* 303–304). This means, in his presentation as in Christian doctrine, that there is no "compellingly" active, physical necessity for the raising of the dead, but there is an appropriateness in relation to the nature of the omega point as creator. This argumentation could be strengthened by the thought that creation is itself an expression of the free love of God, which grants existence to creatures. The creation of the universe and its eschatological fulfillment in the resurrection of the dead may be traced to the same motive for divine action.

In concluding my reflections on Tipler's eschatology, I must address the connection between the Christian hope for the resurrection of the dead and faith in Jesus' resurrection. For Christian believers, communion with Jesus, the crucified and risen One, guarantees participation in the future resurrection of the dead. Tipler did not address this in his lecture at Innsbruck, but he writes in a section of his book entitled "Why I Am Not a Christian" (*Physics* 374ff.) that he cannot believe in the resurrection of Jesus, for historical reasons. It is certainly remarkable that the historians and exegetes who do not regard the resurrection of Jesus as a fact appeal to the natural sciences for support because these supposedly exclude the possibility of such an event. That is not the case for Tipler, since he actually gives physical reasons for the possibility of an eschatological raising of the dead. Should, then, the possibility of such an event within history be discussed again from that new point of view? The event at the universe's end is, after all, according to Tipler's statements, linked to present life and not merely counter to it. May it not be possible, then, that, in accordance with the immanence of the transcendent God, what is eschatologically final could also be present in the midst of history? As regards the historical question, the opinions of many exegetes are in agreement that the core of the Christian Easter traditions is not a matter of legends, and if the content of what was reported were not so unusual, there would be no doubt of its historicity. The stumbling block is the supposed physical impossibility, and because of that, we have alternative reconstructions of the tradition that are historically more improbable than the central assertions of the Christian tradition.

Tipler thinks that he would see the matter differently "if the appearance of such a person at a particular stage of human history were necessary for the omega point to be attained at the end" (*Physics* 380). According to Christian teaching that is, in fact, the case, because human beings estranged from God require the restoration of their communion with God in order that the light of divine eternity may not one day be for them a consuming fire. The sending of Jesus served to accomplish such a restoration of communion with God, whom Jesus, as the "Son" of the Father embodied in his own person, and according to the primitive Christian witness, his mission was attested through his resurrection.

# A Modern Cosmology

Since the risen Christ, according to Christian teaching, through his exaltation, is already participating in God's rule over the cosmos, Christians do not require, in order to hope in the resurrection, the wearisome path by way of the advance of intelligent life to the computer, which will ultimately rule the universe. Communion with the crucified and risen Christ, who according to Christian faith has already been exalted to participation in God's rule of the universe, is a sufficient basis for their hope in their own future participation in the resurrection of the dead. This does not exclude the possibility that the development of life in the universe might actually take the course Tipler describes. But it is precisely the premises of Christology that show that Christian theology cannot simply resolve itself into Tipler's model; rather, it can regard that model only as a theoretical physicist's construct approaching the object of Christian theology. Nevertheless, the fact that such an approach could be developed is significant enough.

# Acknowledgments

Permission to reprint from the following sources is acknowledged gratefully:

Chapter 1: Previously unpublished.

Chapter 2: Published in German as "Wie war ist das Reden von Gott?" *Grudlagen der Theologie.* Edited by Sigurd Martin Daecke and Hans Norbert Janowski. Stuttgart: Kohlhammer, 1974, 29–41. Translated for this edition by Linda Maloney.

Chapter 3: Published in German as "Theologie der Schöpfung und Naturwissenschaft," *Natur und Mensch—und die Zukunft der Schöpfung. Beiträge zur systematischen Theologie,* vol. 2. Göttingen: Vandenhoeck & Ruprecht, 2000, 30–42. Translated for this edition by Linda Maloney. Originally written in 1995.

Chapter 4: Published in *Zygon* 41 no. 1 (2006): 105–112.

Chapter 5: Published in *The Whirlwind in Culture: In Honor of Langdon Gilkey.* Edited by Donald W. Musser and Joseph L. Price. Bloomington: Meyer-Stone, 1977, 171–182.

Chapter 6: Published in *Zygon* 36, no. 4 (2001): 783–784.

Chapter 7: Published in German as "Religion und menschliche Natur," *Natur und Mensch—und die Zukunft der Schöpfung. Beiträge zur systematischen Theologie,* vol. 2. Göttingen: Vandenhoeck & Ruprecht, 2000, 260–270. Translated for this edition by Linda Maloney. Originally written in 1986.

Chapter 8: Published in *Science and Theology: The New Consonance.* Edited by Ted Peters. Boulder, CO: Westview, 1988, 137–148 and in *Natur und Mensch—und die Zukunft der Schöpfung. Beiträge zur systematischen Theologie,* vol. 2. Göttingen: Vandenhoeck & Ruprecht, 2000, 112–122. Reprinted by permission of Westview Press, a member of Perseus Books Group.

Chapter 9: Published in German as "Bewusstein und Geist," *Natur und*

*Mensch—und die Zukunft der Schöpfung. Beiträge zur systematischen Theologie*, vol. 2. Göttingen: Vandenhoeck & Ruprecht, 2000, 123–140. Translated for this edition by Linda Maloney. Originally written in 1983.

Chapter 10: Published in German as "Der Mensch als Person," *Natur und Mensch—und die Zukunft der Schöpfung. Beiträge zur systematischen Theologie*, vol. 2. Göttingen: Vandenhoeck & Ruprecht, 2000, 162–169. Translated for this edition by Linda Maloney. Originally written in 1986.

Chapter 11: Published in German as "Agression und die theologische Lehre von der Sünde," *Natur und Mensch—und die Zukunft der Schöpfung. Beiträge zur systematischen Theologie*, vol. 2. Göttingen: Vandenhoeck & Ruprecht, 2000, 220–234. Translated for this edition by Linda Maloney. Originally written in 1977.

Chapter 12: Published in *Metaphysics and the Idea of God*. Translated by Philip Clayton. Grand Rapids: Eerdmans, 1988, 150–170.

Chapter 13: Previously unpublished.

Chapter 14: Published in *Metaphysics and the Idea of God*. Translated by Philip Clayton. Grand Rapids: Eerdmans, 1988, 113–129.

Chapter 15: Published in *John B. Cobb's Theology in Process*. Edited by David Ray Griffin and Thomas J. J. Altizer. Translated by David Polk. Philadelphia: Westminster Press, 1977, 133–149.

Chapter 16: Published in German as "Eine moderne Kosmologie: Gott und die Auferstehung der Toten," *Natur und Mensch—und die Zukunft der Schöpfung. Beiträge zur systematischen Theologie*, vol. 2. Göttingen: Vandenhoeck & Ruprecht, 2000, 93–98. Translated for this edition by Linda Maloney.

# Notes

1. Later published as "Wie ist eine evangelische Theologie als Wissenschaft möglich" [How Is It Possible for an Evangelical Theology to Be a Science?], *Zwischen den Zeiten* 9 (1931): 8–35.

2. Karl Barth, *Church Dogmatics,* trans. G. T. Thompson et al., ed. G. W. Bromily and T. F. Torrance (Edinburgh: T & T Clark, 1936–1977), I/1, 8.

3. Karl Barth, "Die Theologie und der heutige Mensch" [Theology and the Modern Man], *Zwischen den Zeiten* 8 (1930): 384.

4. Editor's note: The German word is *Religionswissenschaft*. It should be noted, however, that the German term "Wissenschaft" is broader than the English "science," since the German term includes the human and social sciences, hence also what is usually termed "religious studies."

1. Karl Barth, *Church Dogmatics,* trans. G. T. Thompson et al., ed. G. W. Bromily and T. F. Torrance (Edinburgh: T. & T. Clark, 1936–1977), III/1, ix.

2. Karl Heim, *Christian Faith and Natural Science* (London: SCM Press, 1953).

3. Charles Gore, ed., *Lux Mundi: A Series of Studies in the Religion of the Incarnation* (London: J. Murray, 1889).

4. Frank Tipler, *The Physics of Immortality: Modern Cosmology, God, and the Resurrection of the Dead* (New York: Doubleday, 1994).

5. Wolfhart Pannenberg, "Kontingenz und Naturgesetz," in A. M. Klaus Müller and Wolfhart Pannenberg, *Erwägungen zu einer Theologie der Natur* (Gütersloh: Gerd Mohn, 1970), 33–80, translated as "Contingency and Natural Law" in Wolfhart Pannenberg, *Toward a Theology of Nature: Essays on Science and Faith*, ed. Ted Peters (Louisville, Ky.: Westminster/John Knox Press, 1993), 72–122.

CHAPTER 4: PROBLEMS BETWEEN SCIENCE AND THEOLOGY
IN THE COURSE OF THEIR MODERN HISTORY

1. Andrew Dickson White, *A History of the Warfare of Science with Theology in Christendom* (New York: D. Appleton, 1896).

2. Edwin A. Burtt, *The Metaphysical Foundations of Modern Science*, 2nd ed. (New York: Doubleday Anchor Books, 1932), 113.

3. Ibid.

4. See A. Koyré, *Newtonian Studies* (London: Chapman and Hall, 1965), 93–94, 109.

5. Ibid., 109.

6. Burtt, *The Metaphysical Foundations*, 243.

7. Ibid., 261.

8. See Max Jammer, *Concepts of Force: A Study in the Foundations of Dynamics* (Cambridge, Mass.: Harvard University Press, 1957), 188ff., 200ff.

9. See Mary B. Hesse, *Forces and Fields: The Concept of Action at a Distance in the History of Physics* (London: Nelson, 1961), 201ff.

10. See Max Jammer, "Feld," in *Historisches Wörterbuch der Philosophie* 2 (1972): 924.

11. See William Berkson, *Fields of Force: The Development of a World View from Faraday to Einstein* (New York: Wiley, 1974), esp. 50ff., 148ff., and 317–318.

12. Ian Barbour, *When Science Meets Religion* (New York and London: SPCK, 2000), 28–29.

13. Jacques Monod, *Chance and Necessity* (New York: Knopf, 1972).

14. Thomas Torrance, *Divine and Contingent Order* (New York: Oxford University Press, 1981).

15. Robert J. Russell, "Contingency in Physics and Cosmology," *Zygon* 23 (1988): 23–43.

16. Hans-Peter Dürr, "Über die Notwendigkeit, in offenen Systemen zu denken," in *Die Welt als offenes System*, ed. G. Altner (Frankfurt: Fischer, 1986), 17.

17. Ilya Prigogine, *From Being to Becoming: Time and Complexity in the Physical Sciences* (San Francisco: W. H. Freeman & Company, 1980).

18. Thus, we have the title of a book by the biochemist Sjoerd L. Bonting, *Chaos Theology: A Revised Creation Theology* (Ottawa: Novalis Publishing, 2002). See also A. Ganoczy, *Chaos, Zufall, Schöpfungsglaube* (Mainz: Griinewald, 1995). See also R. J. Russell, Nancey Murphy, and Arthur R. Peacocke, eds., *Chaos and Complexity: Scientific Perspectives on Divine Action* (Vatican City State: Vatican Observatory, 1995).

19. Bonting, *Chaos Theology*, 33.

20. Ibid., 51.

21. Ibid.

CHAPTER 5: PROVIDENCE, GOD, AND ESCHATOLOGY

1. Langdon Gilkey, *Reaping the Whirlwind: A Christian Interpretation of History* (New York: Crossroad, 1981), 266.

2. Ibid., 216ff., especially 225–226.

3. Ibid., 226, 236–237, 258.

4. Ibid., 231.

5. Ibid., 229ff. Compare my *Theology and the Kingdom of God* (Philadelphia: Westminster Press, 1969), 51ff., especially 55ff.

6. Gilkey, *Reaping the Whirlwind*, 229.

7. Pannenberg, *Theology and the Kingdom of God*, 61; compare, 63. See also the critical remark on p. 126 addressing the tendency of modern revolutions to disdain and destroy the values of the past, together with what is termed there conversion "to the present in the hope of fulfillment."

8. Gilkey, *Reaping the Whirlwind*, 247.

9. See Wolfhart Pannenberg, *What Is Man?* (Philadelphia: Westminster Press, 1970), and Pannenberg, *Anthropology in Theological Perspective* (Philadelphia: Westminster Press, 1985).

10. Gilkey, *Reaping the Whirlwind*, 234.

11. Ibid., 219ff.

12. Ibid., 236–237.

13. Compare especially my little book *Human Nature, Election, and History* (Philadelphia: Westminster Press, 1977) and the critical discussion of liberation theology in my *Christian Spirituality* (Philadelphia: Westminster Press, 1983), particularly chap. 3, "Sanctification and Politics."

14. The first chapter of *Theology and the Kingdom of God* alone would be enough to document the extent of my agreement with Gilkey on this requirement as he formulates it in *Reaping the Whirlwind*, 135–136, and on many subsequent occasions. I only wonder why he thinks that I did not explicitly draw this consequence. I always thought I had done so since 1959. However, I never found it possible to adopt some metaphysical design as it is offered in the philosophical tradition or by contemporary philosophers, for the purpose of theology, because none of them meets the specific requirement for a Christian doctrine of God as it has to be developed today.

15. Gilkey, *Reaping the Whirlwind*, 249. Creativity must not be treated as a separate ontological principle, as in Whitehead, but "as the divine power of being, as the activity of God as creator and preserver" (414, n.34).

16. By mentioning the process thinkers, Heidegger and Bloch side by side, I do not want to suggest that the level of conceptual analysis in those cases is comparable. In the case of Heidegger on the one hand and the process thinkers on the other, there is a considerable degree of conceptual rigor, although in quite different ways; there is much less in the more imaginative language of Bloch. In distinction from Moltmann, I have never been as deeply influenced by Bloch's thought as Gilkey in some places seems to assume.

17. Gilkey, *Reaping the Whirlwind*, 199–200; compare 188ff.

18. Ibid., 200, defines temporal passage as "the movement of events from possibility to actuality."

19. "If entities and the event in which they participate in time are *self-creative*, then temporal passage becomes the prime locus of being and the ground of activity" (ibid., 200).

20. Here my judgment differs from Gilkey's as presented in *Reaping the Whirlwind*, 168 and 384, n.40.

21. Gilkey, *Reaping the Whirlwind*, 234–235.

22. On this question, see my *Human Nature, Election, and History*, especially chaps. 3 and 5.

23. Gilkey, *Reaping the Whirlwind*, 250ff.

24. Ibid., 49.

### CHAPTER 6: A DIALOGUE

1. Robert J. Russell, "Contingency in Physics and Cosmology: A Critique of the Theology of Wolfhart Pannenberg," *Zygon* 23 (March 1988): 23–43.

2. This does not exclude all forms of teleology from the description of physical process. My own way of speaking of God as the power of the future and of drawing the entire process of the universe (and especially human history) toward himself involves some form of teleology. But here, the *telos* is transcendent, while the teleology criticized in the text is concerned with some implanted *telos*, like an intrinsic force directing a process toward its goal.

3. Max Jammer, "Art, Feld, Feldtheorie," *Historisches Wörterbuch de Philosophie* 2 (1972): 923–926.

4. John Polkinghorne, "Wolfhart Pannenberg's Engagement with the Natural Sciences," *Zygon* 34 (March 1999): 154. See also Polkinghorne's shorter remarks in *Belief in God in an Age of Science* (New Haven, Conn.: Yale University Press, 1998), 82.

5. Georg Süssmann, "Geist und Materie," in *Gott—Geist—Materie. Theologie und Naturwissenschaft im Gespräch*, ed. H. Dietzelbinger and Lutz Mohaupt (Hamburg, Germany: Lutherisches Verlagshaus, 1980), 14–31.

6. Albert Einstein, "Die Grundlage der allgemeinen Relativitätstheorie," in *Das Relativitatsprinzip*, ed. H. Lorentz, A. Einstein, and H. Minkowski, 5th ed. (1913; Darmstadt, Germany: Wissenschaftliche Buchgesellschaft, 1958), 108–109.

7. See Max Jammer, *Concepts of Force: A Study in the Foundation of Dynamics* (Cambridge, Mass.: Harvard University Press, 1957), 158–187.

8. Ibid., 201. See also Jammer's article on field concept, above n. 3.

9. See Polkinghorne, "Wolfhart Pannenberg's Engagement with the Natural Sciences," 154; and J. Wicken, "Theology and Science in the Evolving Cosmos: A Need for Dialogue," *Zygon* 23 (March 1988): 48.

10. This is not to deny the analogical use of language in physics as elsewhere. Mark Worthing is correct in pointing to "the analogical character of field concept, especially within quantum physics" (Mark Worthing, *God, Creation, and Contemporary Physics* [Minneapolis: Fortress Press, 1996], 118-119). The application of the field concept to physics involves metaphor and so does its adaptation for theological use. But there is a difference between a vague analogy and a linguistic transfer that by definition and argument constitutes a new conceptual use.

11. With regard to Kant's concept of time, compare Karl H. Manzke, *Ewigkeit und Zeitlichkeit. Aspekte für theologische Deutung der Zeit* (Göttingen, Germany: Vandenhoeck und Ruprecht, 1992), 151ff. Originally, Kant shared the theological interpretation for the givenness of time as infinite in our intuition (82ff.), but later he replaced his idea with an anthropocentric interpretation. This remains implausible, however, since the human subject and self-consciousness, which is finite, can hardly guarantee the objective validity of an infinite totality of time and space (153).

12. For the impact of relativity on the philosophy of time, see William L. Craig, *Time and Eternity: Exploring God's Relationship to Time* (Wheaton, Ill.: Crossway Books, 2001), chap. 2, II.

13. That is to be emphasized in view of the concern expressed by some that my interpretation is "overly bound to physical science" (Wicken, "Theology and Science in the Evolving Cosmos," 48, 51-52). I fully agree with the excellent discussion of the issue by Mark W. Worthing and with his judgment that "it would be a mistake . . . to build any part of our theology on a specific physical theory" (Worthing, *God, Creation, and Contemporary Physics*, 124). When I observed biblical references to cosmic forces as angels and argued for a possible appropriation of this view in theology (Wolfhart Pannenberg, *Systematic Theology*, vol. 2 [Grand Rapids: Wm. B. Eerdmans, 1994], 102ff.), I did not mean to confuse science and theology but intended to express the old Christian confidence that the transcendent God is present and active in his creation, as his eternity is present and active in temporal events and the undivided infinite space of God's immensity is present in the parts of geometrical space. His presence does not exclude the activity of his creatures but rather works through their special forms of activity.

14. Polkinghorne, "Wolfhart Pannenberg's Engagement with the Natural Sciences," 154.

15. See William Berkson, *Fields of Force: The Development of a World View from Faraday to Einstein* (London: Routledge and K. Paul, 1974), 50-51. See also the observations on the relationship of James Clerk Maxwell to Faraday in T. F. Torrance and James Clerk Maxwell, eds. *A Dynamical Theory of the Electromagnetic Field* (Edinburgh: Scottish Academic Press, 1982), preface, 7-8.

16. Jammer, *Concepts of Force*, 200ff., especially 211ff., 257-258. That could mean that the concept of force is returned to theology, which will continue with the apostle Paul to speak of God as forceful *dynamis* (Rom.

1:20). However, short of the unified field theory that Einstein wanted to establish, scientists continue to speak of four basic natural forces: gravitation, electromagnetism, and strong and weak reciprocal forces.

17. Max Jammer, *Concepts of Mass in Classical and Modern Physics* (Cambridge, Mass.: Harvard University Press, 1961), final chapter.

18. Polkinghorne, "Wolfhart Pannenberg's Engagement with the Natural Sciences," 155.

19. Ilya Prigogine and Isabelle Stengers, *Order Out of Chaos: Man's New Dialog with Nature* (New York: Bantam Books, 1984).

20. For further details, see the excellent description given in Philip Hefner, "The Role of Science in Pannenberg's Theological Thinking," in *The Theology of Wolfhart Pannenberg*, eds. Carl E. Braaten and Philip Clayton (Minneapolis: Augsburg Fortress, 1988), 275ff. See also Pannenberg, *Systematic Theology*, vol. 2, 126ff.

21. Polkinghorne, "Wolfhart Pannenberg's Engagement with the Natural Sciences," 154; compare Niels H. Gregersen, "God's Public Traffic: Holistic versus Physicalist Supervenience," in *The Human Person in Science and Theology*, ed. N. H. Gregersen et al. (Edinburgh: T & T Clark, 2000), 153–188.

22. For details, see Pannenberg, *Systematic Theology*, vol. 2, 76–115, especially "The Cooperation of Son and Spirit in the Work of Creation," 109ff.

CHAPTER 7: RELIGION AND HUMAN NATURE

1. Helmuth Plessner, *Die Frage nach der Conditio humana. Aufsätze zur Philosophischen Anthropologie* (Frankfurt: Suhrkamp, 1976), 7–81, quotation at 67.

2. Karl J. Narr, "Beiträge der Urgeschichte zur Kenntnis der Menschennatur," in *Neue Anthropologie*, eds. Hans-Georg Gadamer and Paul Vogler, Kulturanthropologie 4 (Stuttgart: Thieme; Munich: Deutscher Taschenbuchverlag, 1973), 3–62.

3. Anthony F. C. Wallace, *Religion: An Anthropological View* (New York: Random House, 1966).

4. Elman R. Service, *Origins of the State and Civilization: The Process of Cultural Evolution* (New York: Norton, 1975).

5. Michael Landmann, *Philosophische Anthropologie. Menschliche Selbstdeutung in Geschichte und Gegenwart* (Berlin: de Gruyter, 1955; 4th revised and expanded ed. Berlin and New York, 1976) [English: *Philosophical Anthropology*, trans. David J. Parent (Philadelphia: Westminster, 1974)]; see also Michael Landmann, *Pluralität und Antinomie. Kulturelle Grundlagen seelischer Konflikte* (Munich and Basel: E. Reinhardt, 1963).

6. Jean Piaget, *La formation du symbole chez l'enfant: imitation, jeu et rêve, image et représentation* (Neuchâtel, Switzerland: Delachaux et Niestlé, 1959) [English: *Play, Dreams, and Imitation in Childhood*, trans. C. Gattegno and F. M. Hodgson (New York: Norton, 1962)].

7. Wolfhart Pannenberg, *Anthropologie in theologischer Perspektive* (Göttingen, Germany: Vandenhoeck & Ruprecht, 1983), 347–348 [English: *Anthropology in Theological Perspective,* trans. Matthew J. O'Connell (Philadelphia: Westminster, 1985)].

8. Ibid., 348.

9. Peter L. Berger, *A Rumor of Angels: Modern Society and the Rediscovery of the Supernatural* (Garden City, N.Y.: Doubleday, 1969).

10. Peter L. Berger, Brigitte Berger, and Hansfried Kellner, *The Homeless Mind: Modernization and Consciousness* (New York: Random House, 1973).

### CHAPTER 9: CONSCIOUSNESS AND SPIRIT

1. Karl R. Popper and John C. Eccles, *The Self and Its Brain: Argument for Interactionism* (Berlin and New York: Springer International, 1977, 1981), 179–180: "But how could the unextended soul exert anything like a push on an extended body?" (180).

2. John Locke, *An Essay Concerning Human Understanding,* II, 23, 5.

3. David Hume, *A Treatise on Human Nature* (1730/40) I, iv, 5. (See David Hume, *A Treatise of Human Nature,* 2nd ed., annotated by Lewis A. Selby-Bigge and Peter H. Nidditch [Oxford: Clarendon Press; New York: Oxford University Press, 1978], 232ff., 250; compare section 6, 251ff.)

4. Gilbert Ryle, *The Concept of the Mind* (London: Hutchinson, 1949).

5. See n. 1, above.

6. Popper and Eccles, *The Self and Its Brain,* 56ff.

7. Ibid., 22ff., especially 27ff.

8. Ibid., 12–13.

9. Ibid., 73 and 13; compare also 30, and frequently elsewhere.

10. Ibid., 109.

11. Ibid., 115; compare 554ff.

12. Ibid. 437ff., especially 441ff. Eccles spoke much more reservedly about accepting a consciousness in animals, even primates; compare 518–519; 534–535; and n.30.

13. Ibid., 38; compare the remarks on language as a tool at 48–49.

14. Locke, *An Essay Concerning Human Understanding,* II, 23, 18, and 15.

15. Ibid., II, 23, 5.

16. Augustine, *De lib. arb.* I, 8, 18: . . . *hoc quicquid est, quo pecoribus homo praeponitur, sive mens, sive spiritus, sive utrumque rectius appellatur . . . Ratio ista ergo, vel mens, vel spiritus cum irrationales animi motus regit, id scilicet dominatur in homine cui dominatio lege debetur ea, quam aeternam esse comperimus.* Compare *De Trin.* XIV, 16 (CCL 50a, 453, 35ff.). Usually, Augustine prefers to speak of consciousness (*mens*) or reason (*ratio*). This may be connected to the complexity of the word *spiritus* (compare the different meanings of the expression listed in *De Gen. ad litt.* XII, 7–8, and *De Trin.* XIV, 16). But it may also be associated with the danger

of confusing the expression *spirit* with the divine Spirit; on this see *De Gen. c. Manich.* II, 8, and also *De Gen. ad litt.* VII, 2ff. It may be no accident that the anthropological discussions in the work on the Trinity do not refer to Gen. 2:7, with the exception of the fleeting remark at II, 18, 34.

17. Thomas Aquinas, *Summa theol.* I, 97, 3c: ... *anima rationalis et anima est et spiritus. Dicitur autem esse anima secundum illud quod est commune ipsi et aliis animabus, quod est vitam corpori dare ... Sed spiritus dicitur secundum illud quod est proprium ipsi, et non aliis animabus, quod scilicet habeat virtutem intellectivam immaterialem.*

18. Ibid., I, 36, 1c: *nomen spiritus in rebus corporeis impulsionem quandam et motionem significare videtur; nam flatum et ventum spiritum nominamus.* Compare Augustine, *De Trin.* XIV, 16 (CCL 50a, 452, 32ff.).

19. Immanuel Kant, *Anthropologie in pragmatischer Hinsicht* (Königsberg: Friedrich Nicolovius, 1798; English: *Anthropology from a Pragmatic Point of View,* trans. and ed. Robert B. Louden [Cambridge and New York: Cambridge University Press, 2006]), §57; Friedrich W. J. Schelling, "Ideen zu einer Philosophie der Natur" (1797), in *Werke,* ed. Karl F. A. Schelling, 14 vols. (Stuttgart: Cotta, 1856–61), II, 51. See also Odo Marquard, art. "Geist," *Historisches Wörterbuch der Philosophie* 3 (1974): 182–191, at 184–185, 186–187. According to Marquard, this romantic version of the concept of spirit combines motifs from the aesthetic of genius and theology (187–188). For the post-Hegelian reduction of the concept of spirit to individual consciousness, see ibid., 199–200.

20. Walther Zimmerli, *Ezechiel,* 2 vols., BKAT 13 (Neukirchen-Vluyn: Neukirchener Verlag, 1955–1969) 2:895; compare 900 [English: *Ezekiel 2: A Commentary on the Book of the Prophet Ezekiel, Chapters 25–48,* trans. James D. Martin; ed. Paul D. Hanson with Leonard Jay Greenspoon. Hermeneia (Philadelphia: Fortress Press, 1983)]. When Zimmerli (895) distinguishes Ezekiel's concept from Eccl. 12:7 with the note that in the prophet's vision the *ruah* does not come from God but is summoned from the place of its presence in the world, that is a correct description of the vision itself (compare Ezek. 37:9); but Zimmerli's remark does not do full justice to the explanation of the vision in Ezek. 37:14, where the breath of life is explicitly attributed to the Spirit of God.

21. See Wolf-Dieter Hauschild, *Gottes Geist und der Mensch. Studien zur frühchristlichen Pneumatologie* (Munich: Chr. Kaiser, 1972), 30ff.; 36ff. (on Clement); 89ff. (on Origen); 152ff. (on the Gnostics); 201ff. (on Tatian); 206ff. (on Irenaeus).

22. Ibid., 41–42 (on Clement).

23. According to Ernst Cassirer, *Philosophie der symbolischen Formen* (Berlin: B. Cassirer, 1923–1929), 1:56–57 [English: *The Philosophy of Symbolic Forms,* trans. Ralph Mannheim (New Haven, Conn.: Yale University Press, 1953–57)], this is characteristic of mythic speech, whereas the distinction between language and myth results from the reflection that the object is only represented by language but is not itself present (2:53).

24. Jean Piaget, *La formation du symbole chez l'enfant; imitation, jeu et rêve, image et représentation* (Neuchâtel: Delachaux & Niestlé, 1945); English: *Play, Dreams, and Imitation in Childhood*, trans. C. Gattegno and F. M. Hodgson (New York: Norton; London: Heinemann, 1951); German: *Nachahmung, Spiel und Traum* (1975), 116ff., 127ff., 310ff., 316ff.

25. In his lecture, "The Evolution of Language in the Late Pleistocene," at the 1976 meeting of the New York Academy of Sciences (published in *Origins and Evolution of Language and Speech*, ed. Stevan R. Harnad, Horst D. Steklis, and Jane Lancaster, *Annals of the New York Academy of Sciences* 280 [New York: New York Academy of Sciences, 1976], 312–325, especially 319), Julian Jaynes assumes a close reciprocal relationship between the origins of language, art, and religion. Susanne K. Langer suggests that the festival ceremonies of earlier cultic rituals might have furnished the occasion for the first development of language (*Mind: An Essay on Human Feeling* [Baltimore: Johns Hopkins University Press, 1967], 2:303ff., 307–308).

26. Bernhard Rosenkranz, *Der Ursprung der Sprache: Ein linguistisch-anthropologischer Versuch* (Heidelberg: C. Winter, 1961), 112–113; 114ff. Compare the arguments for the contrary position in Ashley Montague, "Toolmaking, Hunting and Language," in Stevan R. Harnad, Horst D. Steklis, and Jane Lancaster, eds., *Origins and Evolution of Language and Speech* (New York: New York Academy of Sciences, 1976), 266–274.

27. Popper and Eccles, *The Self and Its Brain*, 1–35.

28. Georg Süssmann, "Geist und Materie," in Hermann Dietzfelbinger and Lutz Mohaupt, eds., *Gott—Geist—Materie. Theologie und Naturwissenschaft in Gespräch.* Zur Sache, vol. 21 (Hamburg: Lutherisches Verlagshaus, 1980), 14–31, quotation at 20.

29. Ibid., 22–23.

30. See Hans-Walter Wolff, *Anthropologie des Alten Testaments* (Munich: Kaiser, 1973), 25–40 [English: *Anthropology of the Old Testament*, trans. Margaret Kohl (Philadelphia: Fortress Press, 1974)].

31. Like Popper, Eccles also follows Sir Charles Sherrington in thinking that the crucial mark of human consciousness is to be found in its integrative unity (see John C. Eccles and William C. Gibson, *Sherrington, His Life and Thought* [Berlin: Springer Verlag, 1979]; Popper and Eccles, *The Self and Its Brain*, 524, with Popper's remarks at p. 127). But Eccles refers the integrative activity of the mind directly to the diversity of neural cells, modules, and centers in the brain. From his insight that "the unity of conscious experience comes not from an ultimate synthesis in the neural machinery," he concludes that the unity of consciousness must be located "in the integrating action of the self-conscious mind on what it reads out of the immense diversity of neural activities in the liaison brain" (356). According to Eccles, the self-aware mind chooses from among the "multitude of active centers at the highest level of brain activity" (362) and accomplishes the integration of this plurality ("from moment to moment integrates its selection to give unity even to the most transient experiences" (362 and 478,

488). But this opinion stands in contrast to the fact that our mind knows nothing of the "brain events" except at a later stage of human research and neurophysiological science. The integrative activity of the consciousness is applied to momentary perceptions and memories, not to brain processes. The synthesis of these brain processes in momentary perceptions ("a unified conscious experience of a global or Gestalt-character" [362], operates on a different level than the integrative activity of the reflective consciousness. For a distinction between the emerging consciousness evident even in higher animals (birds and mammals) and human self-awareness, see now John C. Eccles, "Animal Consciousness and Human Self-Consciousness," *Experientia* 38 (1982): 1384–1391, especially 1386ff., in regard to Donald R. Griffin, *The Question of Animal Awareness: Evolutionary Continuity of Mental Experience* (New York: Rockefeller University Press, 1976).

### CHAPTER 10: THE HUMAN BEING AS PERSON

1. Michael Theunissen, "Skeptische Betrachtungen über den anthropologischen Personbegriff," in *Die Frage nach dem Menschen: Aufriss einer philosophischen Anthropologie*, ed. Heinrich Rombach (Freiburg and Munich: Alber, 1966).

2. See, for example, Paul Christian, *Das Personverständnis im modernen medizinischen Denken* (Tübingen: J. C. B. Mohr, 1952).

3. Immanuel Kant, *Anthropologie in pragmatischer Hinsicht*, in *Werke*, vol. 7 (Berlin: Reimer, 1798); [English: *Anthropology from a Pragmatic Point of View*, trans. Victor L. Dowdell, ed. Hans H. Rudnick; introduction by Frederick P. Van De Pitte (Carbondale: Southern Illinois University Press, 1978), §1].

4. Søren Kierkegaard, "Die Krankheit zum Tode," (1849) in *Gesammelte Werke*, ed. Emanuel Hirsch, 36 sections in 24 vols. (Düsseldorf: Diederichs, 1951–66), Section 24; English: *Fear and Trembling and the Sickness unto Death*, trans. with introduction and notes by Walter Lowrie (Garden City, N.Y.: Doubleday, 1954)].

### CHAPTER 11: AGGRESSION AND THE
### THEOLOGICAL DOCTRINE OF SIN

1. Konrad Lorenz, *Das sogenannte Böse. Zur Naturgeschichte der Agression* (Vienna: Borotha-Schoeler Verlag, 1963). Further references to this work will be cited in the text. [Editor's note: Hence, the German title of the book was "The So-Called Evil"; English translation: *On Aggression*, trans. Marjorie Kerr Wilson (New York: Harcourt Brace Jovanovich, 1966)].

2. Sigmund Freud, *Das Unbehagen der Kultur* (Vienna: Internationaler Psychoanalytischer Verlag, 1930). Further references to this work will be cited in the text. [English translation: *Civilization and Its Discontents*, trans. Jean Riviere (New York: J. Cape and H. Smith, 1930)].

3. Paul Tillich, *Systematische Theologie* 2 (1958): 62. [English: *System-*

*atic Theology*, 3 vols. (Chicago: University of Chicago Press, 1951–1963), 2: 53–55].

4. Arthur Schopenhauer, *Die Welt als Wille und Vorstellung*, Book 4, §61. [English: *The World as Will and Idea*, trans. R. B. Haldane and J. Kemp, 3 vols. (London: Routledge & Kegan Paul, 1964), 1:429].

5. This is overlooked when theories of original aggressiveness and wildness are categorized "naturally" with the "legends" of "innate depravity" or human original sin (so M. F. Ashley Montagu, "The New Litany of 'Innate Depravity,' or Original Sin Revisited," in M. F. Ashley Montagu, ed., *Man and Aggression* [New York: Oxford University Press, 1968], 3–17).

6. For the priority of *amor sui* over concupiscence, compare Augustine, *De trinitate* 12, 9, 14.

7. Thus, for example, Adolf von Harnack, *Dogmengeschichte* III (1889; 4th ed., Tübingen: J. B. C. Mohr, 1910), 218, n.2. [English: *History of Dogma*, trans. from the 3rd German ed. by Neil Buchanan, 7 vols. in 4 (New York: Dover Publications, 1961)]. For example, Joachim Kahl, *Das Elend des Christentums: oder, Plädoyer für eine Humanität ohne Gott*, mit einer Einführung von Gerhard Szezesny (Reinbek bei Hamburg: Rowohlt, 1968) [English: *The Misery of Christianity: or, A Plea for a Humanity without God;* with a preface by Gerhard Szczesny, trans. N. D. Smith (Harmondsworth: Penguin, 1971)], 49ff., refers sweepingly to a "demonization of sexuality" in Christianity.

8. Friedrich Nietzsche, *Zur Genealogie der Moral* (1887), 2, §16. [English: *On the Genealogy of Morals*, trans. Walter Kaufmann and R. J. Hollingdale (New York: Vintage, 1967)]. Oddly enough, Leo Kofler, *Aggression und Gewissen: Grundlegung einer anthropologischen Erkenntnistheorie* (Munich: Hanser, 1973), who sets up conscience as "regulating individuals' relationships to one another" against the inclination to aggression (57; compare 66), says nothing about Nietzsche's (and Freud's) derivation of ("bad") conscience itself from aggression turned inward.

9. Nietzsche, *Zur Genealogie der Moral*, 2, §19.

10. Ibid., 2, §20.

11. Augustine, *De civitate Dei* I, 17.

12. Ibid., XIV, 11, 2: *Postea veroquam superbus ille angelus, ac per hoc invidus, per eandem superbiam a Deo ad semetipsum conversus, quodam quasi tyrannico fastu gaudere subditis quam esse subditus eligens, de spiritali paradiso cecidit . . . malesuada versutia in hominis primi sensus serpere affectans, cui utique stanti quoniam ipse ceciderat, invidebat . . . sermocinatus est feminae. . . .* For the destructive intent of this attitude, see also XI, 13, where the seduction of humanity is characterized as murder (that is, of the soul).

13. Helmut Schoeck, *Der Neid. Eine Theorie der Gesellschaft* (2nd ed.; Freiburg: Herder, 1968), 14–15; compare 90ff.; 118ff., and frequently. [English: *Envy: A Theory of Social Behaviour*, trans. Michael Glenny and Betty Ross (Indianapolis: Liberty Press, 1987)].

14. Augustine, *De civitate Dei*, XII, 3.

15. On this see Rolf Denker, *Angst und Aggression* (Stuttgart: Kohlhammer, 1974).

16. John Dollard, Neal E. Miller, Leonard W. Doob, O. H. Mowrer, Robert S. Sears et al., *Frustration and Aggression* (New Haven, Conn.: Yale University Press for the Institute of Human Relations, 1939).

17. Thus, Denker, *Angst und Aggression*, 89; compare 37. This narrowing of the concept of fear is probably explained by the fact that Denker ignores Kierkegaard's distinction between indeterminate, generalized *Angst* (anxiety) and concrete *Furcht* (fear) directed at a particular object (28). That fear cannot be completely separated from *Angst* because the threatening object always actualizes the primeval existential *Angst* in one way or another is correct, but that does not preclude that *Angst* precedes every particular relation to an object. Thus, while there is *Angst* without a particular relationship to an object, and therefore without fear of the threatening object, the opposite is not true: There is no fear without accompanying *Angst*.

18. Denker, *Angst und Aggression*, 30ff., has developed these connections, following Freud, and solidified the model thus derived through dialogue with other positions in aggression research.

19. Denker's thesis that even the instances adduced by psychology of learning of direct imitation of aggressive behavior not provoked by a frustration are grounded in *Angst* (57–58) can probably be maintained only in terms of the general existential *Angst* that precedes concrete frustration.

20. Friedrich Schleiermacher, *Der christliche Glaube nach den Grundsätzen der evangelischen Kirche im Zusammenhange dargestellt* (Berlin: G. Reimer, 1821–1822), § 72, 2. [Editor's note: Compare the English translation, based on the second German edition (1830), *The Christian Faith*, ed. and trans. H. R. Mackintosh and J. S. Stewart (Edinburgh: T & T Clark, 1989), 292–295].

21. Søren Kierkegaard, *Samlede Vaerker* [1st edition = *SV1*], ed. J. L. Heiberg (Copenhagen: Gyldendal 1901–1906), IV, 320; German translation by Emanuel Hirsch in his edition, *Gesammelte Werke*. 36 Sections in 26 vols. (Düsseldorf: E. Diederich, 1951–1966), Section 11, 48. [English trans. in *The Concept of Anxiety: A Simple Psychologically Orienting Deliberation on the Dogmatic Issue of Hereditary Sin* (Kierkegaard's Writings, VIII), ed. and trans. Reidar Thomte in collaboration with Albert B. Anderson (Princeton, N.J.: Princeton University Press, 1980), 49–50].

22. *SV1*, IV, 315 = Hirsch, Section 11, 42 [*The Concept of Anxiety*, 43–44].

23. The "concept of *Angst*" is first of all about a synthesis of body and soul through the spirit (*SV1*, IV, 315 = Hirsch Section. 11, 42), but this is only a special form of the "synthesis of the infinite and the finite" that is the subject of the "sickness unto death" (*SV1*, XI, 127 = Hirsch Section 24, 60–61). See *The Concept of Anxiety*, 43–44, and *The Sickness unto Death: A Christian Psychological Exposition for Upbuilding and Awakening* and

Kierkegaard's Writings. XIX, eds. and trans. Howard V. Hong and Edna H. Hong (Princeton, N.J.: Princeton University Press, 1980), 13.

24. *SVi*, IV, 331 = Hirsch Section 11, 60–61.

25. Paul Tillich, *Systematische Theologie* (1958), 2:35ff., especially 39ff. [English: *Systematic Theology* 2:44–59]. For Tillich's interpretation of *Angst* in its relationship to psychoanalysis, see Helmut Elsässer, *Paul Tillichs Lehre vom Menschen als Gespräch mit der Tiefenpsychologie* (Stuttgart: Sprint-Druck, 1973), 38ff., 50ff., 103–104, and especially 116ff. This was the author's Ph.D. dissertation from Marburg, 1973.

26. Martin Heidegger, *Sein und Zeit* (Tübingen Max Niemeyer Verlag 1927), 182ff., 191ff. [English: *Being and Time*, trans. Joan Stambaugh (Albany, N.Y.: SUNY Press, 1996)].

27. Rudolf Bultmann, *Theologie des Neuen Testaments* (1953), 237, 239–40 (for the relationship between worry and anxiety) [English: *Theology of the New Testament*, trans. Kendrick Grobel, 2 vols. (New York: Scribner, 1951–1955), 240, 243, 247]. On this subject, compare also Jesus' sayings against worry in Matt. 6:25ff. (Luke 12:22ff.).

28. Alfred Adler, "Der Aggresionstrieb im Leben und in der Neurose" (1908) in *Kritische Studienausgabe*, ed. Almuth Bruder-Bezzel (Munich: DGIP, 1989), part 1, vol. 2, *Aufsätze 1908/1911*.

29. Helmut Nolte, "Über Aggression," in Wolf Lepenies and Helmut Nolte, *Kritik der Anthropologie: Marx und Freud, Gehlen und Habermas über Aggression* (Munich: Hanser, 1971), 103–140, quotation at 125.

30. Ibid., 131.

31. Walter Neidhart, "Was erzeugt Aggression? Hypothesen der Forschung aus theologischer Sicht," *Lutherische Monatshefte* 13 (1974): 418–424, quotation at 422.

32. See ibid., 422–423.

33. This form of Old Testament thinking was first pointed out by Klaus Koch in his essay, "Gibt es ein Vergeltungsdogma im Alten Testament?" *Zeitschrift für Theologie und Kirche* 52 (1955): 1ff. See also Gerhard von Rad, *Theologie des Alten Testaments* (1957), 1:382 [English: *Old Testament Theology*, trans. D. M. G. Stalker, 2 vols. (New York: Harper, 1962–1965), 1:385].

### CHAPTER 12: MEANING, RELIGION,
### AND THE QUESTION OF GOD

1. Paul Tillich, "The Philosophy of Religion," in *What Is Religion?* ed. James Luther Adams (New York: Harper & Row, 1969), chap. 1 [German edition: *Religionsphilosophie* (1925); republished in an Urban-Reihe edition, vol. 63 (1962), 42, 44ff.].

2. Viktor E. Frankl, *The Will to Meaning: Foundations and Applications of Logotherapy* (New York: New American Library, Plume Books, 1969), 156 [German edition: *Der Wille zum Sinn* (1972), 117]. The book stems

from a series of lectures at the Perkins School of Theology, Southern Methodist University, Dallas, Texas, summer 1966.

3. Wilhelm Dilthey, *Gesammelte Schriften*, ed. Bernhard Groethuysen (Leipzig: Tübner, and Göttingen: Vandenhoeck & Ruprecht, 1914–1972), 7:237.

4. Tillich, "The Philosophy of Religion," 59 (= *Religionsphilosophie*, 44).

5. See Wolfhart Pannenberg, *Theology and the Philosophy of Science*, trans. Francis McDonagh (Philadelphia: Westminster, 1976).

6. See Gerhard Sauter, *Was heisst nach Sinn fragen? Eine theologisch-philosophische Orientierung* (Munich: C. Kaiser, 1982), 145, 163.

7. Ibid., 39ff., 46ff., 56ff., 130–131.

8. Ibid., 61–62, 88.

9. For a critique of this position, see ibid., 105, 107–108.

10. Sauter comes at least very close to such a thesis, inasmuch as he flatly characterizes the meaning question (as a question concerning absolute meaning) as "immoderate and presumptuous" (ibid., 167). He brings the alleged avoidance of this question by Job and Kohelet (Ecclesiastes) into connection with the Old Testament prohibitions of images of God. Nevertheless, Sauter also says that the question of meaning belongs to life itself (128–129), and speaks of the "meaning that is communicated in the cross of Christ" (152ff.).

### CHAPTER 13: ETERNITY, TIME, AND SPACE

1. See Wolfhart Pannenberg, "God and Nature," in *Toward a Theology of Nature: Essays on Science and Faith*, ed. Ted Peters (Louisville, Ky.: Westminster, 1993), 50–71, esp. 61–62 and 70, n.51.

2. William Lane Craig, *Time and Eternity: Exploring God's Relationship to Time* (Wheaton, Ill.: Crossway Books, 2001), 66.

3. Ibid., 178.

4. Max Jammer, *Concepts of Space: The History of Theories of Space in Physics*, foreword by Albert Einstein (Cambridge, Mass.: Harvard University Press, 1954); German translation, *Das Problem des Raumes* (1960), 178–179.

5. Ibid., xiv–xv.

6. T. F. Torrance, *Space, Time and Incarnation* (London: Oxford University Press, 1969), 13ff.; compare 7–8.

7. Thus, ibid., 63.

8. See W. L. Craig, *The Tenseless Theory of Time: A Critical Examination* (London-Dordrecht: Kluwer, 2000), 105ff.

### CHAPTER 14: ATOMISM, DURATION, FORM

1. More precisely, "actual occasion" designates the primary constituents of events: "an actual occasion is the limiting type of an event with only one

member." Alfred North Whitehead, *Process and Reality: An Essay in Cosmology,* corrected edition, edited by David Ray Griffin and Donald W. Sherburne (New York: Free Press, 1978), 73; *Process and Reality: An Essay in Cosmology* (New York: Macmillan, 1929), 113. [Below both editions are referred to.]

2. Ibid., 18 (27).

3. Ibid., 35 (53), 27 (40), respectively.

4. Ibid., 67 (104).

5. Ibid., 35 (53).

6. Ibid., 321 (489–490).

7. Samuel Alexander, *Space, Time and Deity* (1920; New York: Macmillan, 1966), 1:143; compare 149.

8. Ibid., 1:48–49.

9. Ibid., 1:45.

10. Ibid., 1:42.

11. Ibid., 1:149.

12. Ibid., 1:321.

13. Ibid., 1:325.

14. Ibid., 1:339ff.

15. Ibid., 1:338ff.

16. Ibid., 1:321.

17. Plato, *Parmenides,* 165ff.

18. A. N. Whitehead, *Adventure of Ideas* (New York: Macmillan Free Press, 1933), 125ff.

19. Ibid., 127.

20. Ibid., 121ff.

21. A. N. Whitehead, *Science and the Modern World,* 2nd ed. (New York: Macmillan Free Press, 1925), 115.

22. Ibid., 98.

23. Ibid., 115.

24. Whitehead, *Process and Reality,* 22 (32). The relations' fading into the background may be connected with the fact that relations may now be characterized as reciprocal ("the complex of mutual prehensions," *Process and Reality,* 194 [295]), while *Science and the Modern World* still distinguishes, in an Aristotelian sense, between internal and external relations (analogous to the difference between *relatio realis* and *relatio rationis* in Scholastic philosophy). Compare also *Process and Reality,* 222–223 (340) and 50 (79).

25. Whitehead, *Process and Reality,* 45 (71–72). Compare. 80 (123): "each actual entity is a locus for the universe," a view expressed earlier in *The Concept of Nature* (Cambridge: Cambridge University Press, 1920), 152. Later Whitehead relates this idea to the concept of a physical field (*Process and Reality,* 80 [123–124]).

26. Whitehead, *Process and Reality,* 80 (124).

27. Whitehead, *Science and the Modern World,* 68.

28. Compare *Adventure of Ideas*, 138.

29. Compare *The Concept of Nature*, 78.

30. Whitehead, *Science and the Modern World*, 97–98.

31. Whitehead, *Process and Reality*, 227 (347).

32. Ibid., 26 (39), 283 (434), respectively.

33. Compare, for example, ibid., 26–27 (40), 248–249 (380–381).

34. Ibid., 86 (131), 88 (135); compare 25–26 (38).

35. Whitehead, *The Concept of Nature*, 190. In this work, Whitehead still did not view the "point-flash" or "event-particle" as the ultimate real component of the natural world: "You must not think of the world as ultimately built up of event particles," he expressly says there (172; compare 59). The world is rather "a continuous stream of occurrences which we can discriminate into finite events forming by their overlappings and containings of each other and separations a spatio-temporal structure" (172–173; compare also 161).

36. Whitehead, *Process and Reality*, 5 (8), 7 (11); compare 4ff. (7ff.).

37. Whitehead, *Science and the Modern World* 129–130.

38. Whitehead, *Process and Reality*, 8 (12), 17 (25), respectively.

39. Compare *Process and Reality* 278 (424–425), 214–215 (327–328); *Adventure of Ideas*, 194–195.

40. So, especially, John Cobb, *God and the World* (Philadelphia: Westminster, 1965), 42–66. God, who encounters humans as their future by calling them to new possibilities, is, according to Cobb, a force (59), indeed a liberating force (64), but also at the same time the ground of our being and of the order of nature (65). Can such statements about a creative work of God justifiably fit into the conceptual scheme of Whitehead's philosophy? Don't they rather demand a reworking of the fundamental categories of his system? According to the remarks on "God and the World" in *Process and Reality* (342–351 [519–533]), God is not creator of the world (346 [526]); or, in any case, he is just as much its creature as its creator (348 [528]). How this conception differs from the Christian view of Creation has been accurately portrayed by Langdon Gilkey in *Reaping the Whirlwind: A Christian Interpretation of History* (New York: Harper & Row, 1977), 248ff. He asserts correctly that Whitehead's thought on philosophy "must be reinterpreted and in part refashioned" (114). But has Gilkey taken sufficient account of the thoroughly radical changes demanded by such a view? He finds Whitehead's concept of "creativity" obscure and even incoherent (250) because in *Process and Reality*, 21 (31–32), it is characterized as the principle which brings forth new entities, while on p. 18 (27) only events, as actual entities themselves, form the ultimately real elements which compose the world (Gilkey, *Reaping*, 414 n.34). Perhaps this difficulty is resolved by viewing the principle of creativity as expressing nothing more than the self-constitution of the events in their "subjectivity." Still, Gilkey feels that the incoherence in Whitehead's conception cannot be overcome "unless it is reinterpreted as the divine power of being, as the activity of God as cre-

ator preserver" (414 n.34). The question then is whether Gilkey, along with Whitehead, can still talk of the self-creativity or "self-actualization" of the world or events, without sounding incoherent himself. A much deeper incursion into Whitehead's conceptual scheme seems necessary if enough room is to be cleared for the idea of a creator God. The idea of the self-constitutive subjectivity of the event itself would need to be altered, and since this idea is closely related to Whitehead's event atomism, both would have to be revised together. Of course, such a correction cannot be based upon some theological postulate. Its necessity can only be shown through philosophical argument.

### CHAPTER 15: A LIBERAL *LOGOS* CHRISTOLOGY

1. Wolfhart Pannenberg, *Jesus—God and Man* (Philadelphia: Westminster Press, 1968), 166–167.

2. Ibid., 168.

3. Ibid., 168–169.

4. John B. Cobb, *Christ in a Pluralistic Age* (Louisville, Ky.: Westminster John Knox Press, 1976), 130ff. Further references to this work will be cited in the text as *CPA*.

5. Alfred North Whitehead, *Process and Reality: An Essay in Cosmology,* corrected edition, edited by David Ray Griffin and Donald W. Sherburne (New York: Free Press, 1978), 350; *Process and Reality: An Essay in Cosmology* (New York: Macmillan, 1929), 532.

6. John B. Cobb, *God and the World* (Philadelphia: Westminster Press, 1969), 84. Further references to this work will be cited in the text as *GW*. Compare Alfred North Whitehead, *Adventures of Ideas* (New York: Macmillan Free Press, 1933), chaps. 19 and 20.

7. Whitehead, *Adventures of Ideas,* 226. "We must conceive the Divine Eros as the active entertainment of all ideals, with the urge to their finite realization, each in its due season. Thus a process must be inherent in God's nature, whereby his infinity is acquiring realization."

8. Whitehead, *Process and Reality,* 346 (526): "He does not create the world, he saves it." Certainly this rejection of the notion of creation is directed against a traditional interpretation of creation which thought of creation as a completed, primordial act. Whitehead's antithesis would lift up the fact that precisely the process of redemption, in its own right, is to be thought of as creation. But Whitehead himself has not done that, and such a notion is also not feasible apart from a thoroughgoing revision of his philosophical principles—because Whitehead conceives of God and world in mutual interdependence (ibid., 347ff. [527ff.]), and that is unifiable with no concept of creation.

9. Whitehead, *Process and Reality,* 343 (521): "It does not look to the future." For a different view, see Cobb, *Christ in a Pluralistic Age,* 85ff.

10. Whitehead, *Adventures of Ideas,* 268; compare 194–195. But

according to Whitehead, the character of every occurrence as an anticipation of the future is not an expression of a constitutive significance of the future for the present, but to the contrary is an expression of the necessity which the present occurrence imposes upon the subsequent future.

11. Compare Cobb, *God and the World*, 3ff. Moreover, Cobb himself remarks there: "The intense focus of attention upon the future, characteristic of much contemporary theology and bound up with its understanding of Jesus Christ, is foreign to both Whitehead and Bonhoeffer. Neither the God who works slowly and quietly by love nor the God who helps us by his suffering can be readily identified with Pannenberg's Power of the future."

12. Compare my critical remarks in "Future and Unity," *Hope and the Future of Man*, ed. Ewert H. Cousins (Philadelphia: Fortress Press, 1972), 64, 72–73.

13. To be sure, I would prefer to speak of "creative formation." Transformation is only a partial aspect of the intended dynamic. Also the form which preceded the transformation has already been attributed to the working of the *Logos*, and the determination of the relation between form and transformation is precisely what is crucial. Cobb does make a valuable contribution in that regard, however, in that (in connection with Whitehead) he conceives of the emergence of new occurrences as a condition for the realization of the contribution of the past to the ongoing course of events (*CPA* 70). Of course, not all new occurrences have equally such a positive function for that which is. Moreover, the concept of "creative transformation" or "creative formation" is by no means simply identical with the concept of the *Logos*, because it signifies God's creative activity altogether, working together in the Father, Son, and Spirit. To that extent, it is more encompassing than the *Logos* concept, which in turn is to be defined as a partial aspect of creative transformation or creative formation (see below, n.18).

14. My explorations into the concept of a field of energy (in "The Doctrine of Spirit and the Task of a Theology of Nature," *Theology* 75 [1972]: 15), cited by Cobb (*CPA* 253), differ from Cobb's use of this idea insofar as Cobb uses the concept for the saving efficacy proceeding from the historical Jesus (*CPA* 116–117), and indeed in the sense that occurrences have aftereffects for the future (*CPA* 118). However, a field of energy is not a function of occurrences but, to the contrary, the occurrences are functions of the field. Accordingly one would think not of the historical but of the risen Christ, who is identical with the future of the one resurrected for the world, as the field of force which determines the present existence of the Christian—which is probably closer to the meaning of Paul.

15. The wider point of comparison appears to me to be just as problematic also: "In both cases the primary implications of the expectation are to reinforce and undergird as important those actions which would also appear as good from more general considerations" (*CPA* 227). Jesus

argues not from a basically creaturely consciousness but the reverse: Only in the light of God's eschatological future does the meaning even of every-day reality, the meaning of creation, become clear. However, that the con-tent of his message also appears as good under more general points of view is unlimitedly valid only within the context of the Christian tradition. Already Nietzsche and Marxism realized the effect of these important limitations.

16. Co-constituted—for according to Cobb, even in the case of Jesus the appropriation of the initial aim remains an act of human self-actualization: Jesus freely chose to allow his selfhood to be constituted through God (*CPA* 173).

17. On this point, see Pannenberg, *Jesus—God and Man*, 334ff.

18. One should also differentiate the concept of "creative transfor-mation" in the sense of the distinction between Father and *Logos*, which, strictly speaking, already characterizes a collaboration of Father and Son in the act of Creation. It might be considered whether it is not precisely the aspect of the immanence of form into the creatures which is associated with the Son or *Logos* in this event.

19. Compare Pannenberg, *Jesus—God and Man*, 61 and 251–252.

20. In his Christian evaluation of the Buddhist criticism of the self (*CPA* 205ff.), Cobb indeed declares himself rightly against a dissolution of individuality and, instead, for its enlargement (ibid., 219–220) in relation to the community (ibid., 213–214). Only the substantialist interpretation of the self as a subject must be dissolved (ibid., 214). Cobb justly wishes thereby to preserve the interest that has found its expression in the Chris-tian emphasis upon the individual (ibid., 215). However, the question is not only of the interest in the personal responsibility of the individual, but of the love of God bestowed upon the lost individual with eternal love: There-fore in Christianity the individual can no longer be subordinated without reservation to the system of the society.

21. Whether one may perceive in this process a completion of the incar-nation (ibid., 62) is nevertheless doubtful: The history of Christian art pre-sented in connection with Malraux describes a process of the dissolution of Jesus into the *Logos* (or into an efficacy of the risen Christ), whereas the concept of the incarnation characterizes the entering of the *Logos* into the historical figure of Jesus. Also the generalization of the title "Christ" into the designation of every immanence of the *Logos* (compare ibid., 63, 87) requires at least a more exact grounding and differentiation, if the specifi-cally messianic meaning of the title "Christ" is not to disappear.

22. Compare my explanation in "Future and Unity," in Cousins (ed.), *Hope and the Future of Man*, 69ff. Cobb presents my interpretation one-sidedly, in that he discusses it only under the aspect of the resurrection hope in distinction from hope for the Kingdom of Heaven (*Christ in a Pluralistic Age*, 239ff., 250ff.). My adhering to the bodiliness of resurrection—which, alongside the binding of the resurrection of the dead with the end of our

time, forms the major item of Cobb's critique of my interpretation (ibid., 252ff.)—represents in my eyes not so far-reaching an opposition to Cobb's Whitehead-oriented interpretation as the question of the constitutive significance of God's future for the present. If it is sufficiently considered that every actual event remains present in all its aspects to the eternity of God, then the bodiliness of the resurrection, which actualizes this divine memory of temporal events in a creatively new manner, should cause no insurmountable difficulties for understanding.

### CHAPTER 16: A MODERN COSMOLOGY

1. Frank J. Tipler, *Die Physik der Unsterblichkeit. Moderne Kosmologie, Gott und die Auferstehung der Toten* (Gütersloh: Bertelsmann Verlag, 1994) [Original English edition, *The Physics of Immortality: Modern Cosmology, God, and the Resurrection of the Dead* (New York: Doubleday, 1994)]. Further references to this work will be cited in the text as *Physics*.

2. See Thomas Aquinas, *Summa contra Gentiles*, II, 58. And compare Wolfhart Pannenberg, *Systematische Theologie Band III* (Göttingen: Vandenhoeck & Ruprecht, 1993), 620–621 [English translation, *Systematic Theology: Volume 3* (Grand Rapids: William B. Eerdmans, 1998), 575–576].

# Index

absolute space
  contrasted with atomism and
    duration, 176
  contrasted with infinite space,
    164–66, 172
  Newton's view, 64
  relativity theory and, 32–33
action and meaning, 151–52
actual entities, 176, 179–80, 185, 191,
  229n25
actual occasions, 177, 180–86, 228n1
  (ch.14)
Adler, Alfred, 134, 141
Adorno, Theodor W., 17–18
aggression. *See also* sin
  *Angst*, 137–40
  excuse or release, 140–42
  lust and, 131–34
  as a product of Christianity, 133
  relationship to sin, 133–35
  self-aggression, 134–37
  sources and solutions, 142–44
Alexander, Samuel, 33, 176–78
alienation and secularization, 81
Anaximenes, 36
*Angst*
  body and soul, 226n23
  contrasted with fear, 226n17
  contrasted with frustration, 226n19
  sin and, 137–40
animal communication, 79–80
animal consciousness, 221n12
animals, creation of, 90–91
anthropic principle of creation, 63–64,
  205

anthropology and philosophy, 119
anxiety. *See Angst*
Aquinas, Thomas, 3, 43, 106
argument by congruence, 3
Aristotle
  analysis of motion, 184–86
  on contingency, 31
  on the existence of God, 27–28
  on space and time, 172
atomism, 176–79, 185–86
atonement, 50
Augustine
  doctrine of providence, 53
  on eternity, 56, 168
  on lust, 131
  on Satan's aggression, 130–31
  on self-aggression, 136–37
  on sin, 131–33
  on spirit, 106, 221n16
  on time and eternity, 171

"bad conscience," 134–35, 225n8
Barth, Karl
  on creation theology, 25
  critique of, 51–52
  political views, 50
  on Protestant theology as a science,
    11–15
  theocentric theology, 4
Bartley, William, 12
Berger, Peter L., 82–83, 148
Bergson, Henri, 55, 57, 176
Bible as historical documents, 90
Bloch, Ernst, 54, 217n16
body and soul, 101–2, 120, 226n23